Out of My Skull

Out of My Skull

The Psychology of Boredom

JAMES DANCKERT
JOHN D. EASTWOOD

Harvard University Press
CAMBRIDGE, MASSACHUSETTS AND LONDON, ENGLAND 2020

Printed in the United States of America

First printing

Library of Congress Cataloging-in-Publication Data
Names: Danckert, James, 1972- author. | Eastwood, John D. (John David), 1970- author.
Title: Out of my skull : the psychology of boredom / James Danckert and John D. Eastwood.
Description: Cambridge, Massachusetts : Harvard University Press, 2020. | Includes bibliographical references and index.
Identifiers: LCCN 2019045209 | ISBN 9780674984677 (cloth)
Subjects: LCSH: Boredom—Psychological aspects.
Classification: LCC BF575.B67 D343 2020 | DDC 152.4—dc23
LC record available at https://lccn.loc.gov/2019045209

To my brother Paul.
Often bored, never boring
—JD

For Ben and Bronwyn:
may your days be full of
meaningful engagement
—JDE

CONTENTS

Out of My
Skull

INTRODUCTION

• • •

You arrive early to renew your driver's license, hoping to avoid the crowds. No such luck. It seems others must have had the same idea, since the building is teeming with people. Stepping up to the counter you hear: "Please take a number and have a seat."

You shuffle to the nearest chair in dread. Your body twists and fidgets restlessly. You cast your eyes about, looking here and there for relief. After reading all the posters in the room, which outline the regulations for everyone from first-time drivers to crane operators, you slump forward and prop your head in your hands. Ticket numbers are slowly announced. The lethargy comes in waves, punctuated by moments of irritability. You're drained of energy and yet edgy, restless. You read the posters again. The urgency to act builds as time crawls.

Suddenly, you remember your phone. You tremble as you reach for it, anticipating the relief you've been craving. After you insert your earbuds and unlock your phone, your body relaxes, your mind clears, and a soothing calm envelops you. Crisis averted? Perhaps. You've drowned out a very unpleasant feeling. That's a good thing . . . but what if boredom was trying to tell you something?

• • •

Boredom comes for each of us in those moments when we can't see a way forward, when we want to be doing something but don't want to do anything currently on offer. We could call it sluggishness or listlessness. Or, perhaps it's the opposite, a sense of being "antsy," restless for something but unsure of what will satisfy. While boredom can be described in any number of ways, we've all felt it. We contend that we should pay attention to it and understand it. In our view, being bored is quite fascinating, and maybe, just maybe, it might even be helpful.

For years the topic of boredom has been explored by philosophers, historians, and theologians. Yet, despite the fact that boredom is ubiquitous, until now it has received relatively little attention from the scientific community. With *Out of My Skull* we seek to change that trend. Psychology—the scientific study of mind and behavior—is well positioned to shed light on the human experience of boredom. As psychologists who have been publishing research on boredom for the last fifteen years, with expertise spanning neuroscience and clinical psychology, we've developed an understanding of boredom that emphasizes the key concepts of engagement and agency. Our approach has the virtue of being able to account for a wide range of scientific research findings and draws together diverse approaches to boredom.

But despite our conviction that boredom has a message for you, we would not be so bold as to tell you how to live your life. Boredom itself can't tell you what to do, either. In that sense, you are on your own. This is precisely one of boredom's key messages and, by extension, a core theme of this book. As humans, we need self-determined, effective connection with the world. We need to be engaged, mentally occupied, giving expression to our desires and exercising our skills and talents. In short, we have a need for agency. When this need is fulfilled, we flourish. When this need is thwarted, we feel bored, disengaged.

In this regard, boredom reveals an important aspect of being human: we have a strong need to be engaged with the world around us. As we'll see, a number of substitutes for true engagement might be very tempting and might even beat back boredom in the short term. But such temporary salves never last, and boredom will return with a vengeance. After that, it is up to us to embrace our agency.

Boredom is an elusive subject, with tentacles that stretch into diverse areas of human inquiry. In part, this is what makes it so fascinating, but it is also what makes it so maddening. Consider an exchange between Humpty Dumpty and Alice in Lewis Carroll's *Through the Looking-Glass*: Humpty Dumpty "scornfully" asserts to Alice that "When I use a word it means just what I choose it to mean—neither more nor less." Alice correctly responds: "The question is whether you can make words mean so many different things."

The communication breakdown between Alice and Humpty is emblematic of much of boredom inquiry to date. Although committed to the idea that there is no correct or incorrect way to define boredom, we believe more precision is in order.

In *Out of My Skull* we endeavor to define the elusive subject of boredom with a decidedly psychological approach, as we consider boredom to be an experience occurring inside our minds. We further offer an organizing framework for what has been a fragmented field of study. We hope this will function as a common ground for a wide range of readers and scholars to meet and exchange ideas.

We start our journey by asking "What is boredom?" Just as with most everyday experiences, most of us feel we know boredom well—until we try to define it. The closer we look, the more mysterious and intriguing boredom becomes. Next, we ask "What is boredom good for?" Why would evolutionary forces have shaped us to be affected by such a negative experience? As we'll see, it is actually beneficial to have the capacity to be bored. When it does strike, we need not fear. The trick is in responding well to the signal.

Then we turn our attention to the question "What makes us bored?" The answer is not simple. Boredom, like beauty, could be said to be in the eye of the beholder. One person's joy is another's boredom. However, there are key factors within us, as well as within the situations we find ourselves in, that increase our risk of succumbing to boredom. We further examine how boredom involves being cut off from others and from our critical need to create meaning and find purpose. We then scrutinize boredom's opposites to deepen our understanding of the experience and set the stage for considering what optimal responses to boredom might look like.

Boredom is a call to action, a signal to become more engaged. It is a push toward more meaningful and satisfying actions. It forces you to ask a consequential question: What should I do? *Out of My Skull* is not an answer to this question. It's a guide to help you understand boredom's message more clearly.

CHAPTER 1

BOREDOM BY ANY OTHER NAME

• • •

You lift your head from the dishes in the kitchen sink to look out the window into your backyard. An overwhelming sense of restlessness pervades your thoughts. Born of the desire to be doing something, anything . . . but what? Distracted, you hadn't noticed the dog until now.

She's an Australian Shepherd, a speckled grey-blue coat with tan patches framing an alert face. She's used to rounding up sheep or cattle with an uncanny, seemingly innate skill. Two walks a day—and let's be honest, that's a generous estimate—just aren't quite enough for such an active animal. She needs open spaces, activity, a purpose, a job to do. Not so different from you.

Without sheep to corral she contents herself with charting wide arcs at full speed around your lawn. Normally, this brings a smile to your face. Chasing her tail, occasionally catching it, all seems mildly amusing. And pointless, you realize. Just as this dawns on you, she stops her current circuit, catches her breath briefly, and spies you smiling at her. The plaintive look on her face slowly erases your own smile. Your eyes are locked just long enough for her to figure out you're not coming to the rescue. You will do nothing to save her from the tedium. The endless, meaningless running begins again.

> She's bored. You know it, she knows it. And if your dog can be bored, then what hope do you have to fix your own malaise?

• • •

Sir Dedlock: "Is it still raining, my love?"
Lady Dedlock: "Yes, my love. And I am bored to death with it. Bored to death with this place. Bored to death with my life. Bored to death with myself."[1]

That cheery description of Victorian life comes from a TV adaptation of Charles Dickens's appropriately titled novel *Bleak House* in which Dickens introduces the word *boredom* for the first time. Although to be considered a *bore* predates the Dedlocks,[2] and the French had long made use of *ennui*[3] to capture a feeling of listlessness, "boredom" was not in common English usage until late in the nineteenth century.[4] But the lack of an English word to capture the experience doesn't mean boredom didn't already exist.

Boredom, in some form or other, has always been with us; it is a part of our biology shaped by a long evolutionary heritage. Boredom has a complex and fascinating social, philosophical, literary, artistic, and theological history—far too complex to cover in its entirety here.[5] But to truly understand what boredom is, to define it, we must start somewhere.

A Brief History of Boredom

Peter Toohey, in his excellent book *Boredom: A Lively History,* traces the origins of boredom as far back as antiquity.[6] Seneca, a Roman philosopher, may have been the first to write about boredom, linking it to nausea and disgust, driven by the monotony of daily life:

> How long the same things? Surely I will yawn, I will sleep, I will eat, I will be thirsty, I will be cold, I will be hot. Is there no end? But do all things go in a circle? Night overcomes day, day night, summer gives way to autumn, winter presses on autumn, which is checked by spring. All things pass that they may return. I do nothing new, I see nothing new. Sometimes this makes me seasick [nauseous]. There are many who judge living not painful but empty.[7]

Seneca's lament sounds decidedly modern in its complaint of repetition, that there is nothing new under the sun. One could argue that Ecclesiastes gives an earlier description born of a similar lament of monotony. After outlining the attainment of great wealth and prestige, the narrator of Ecclesiastes writes, "Yet when I surveyed all that my hands had done and what I had toiled to achieve, everything was meaningless, a chasing after the wind; nothing was gained under the sun."[8] Both complaints highlight two components of boredom: first, it is a negative experience and second, it feels purposeless, making living seem empty. Toohey even tells us of one Roman village in the second century that memorialized an official who somehow delivered them from intolerable levels of boredom![9]

Boredom, arising from a lack of zest for daily routine, also loomed large in the Middle Ages. Scholars argue that what has come to be called and understood as boredom has roots in the Latin word *acedia,* which referred to a lack of enthusiasm for the spiritual practices that sustained the monastic life—a listless spiritual languor in which rituals such as the burial of the dead lose their significance.[10]

Referred to as the noonday demon,[11] the ceaseless repetition of daily routines led to an odd combination of lethargy and agitation—bedfellows that recur throughout this book—in monks living a cloistered life. Beyond highlighting the oppressive nature of monotony and purposelessness, what both Seneca and the monks show us is that boredom has long been with us—well before Dickens's treatment of it.

It is not until the mid- to late nineteenth century that we start to see explorations of boredom from a psychological viewpoint. As is so often the case in the history of psychology, it was the Germans who got the ball rolling. Theodor Waitz, better known at the time as an anthropologist, and the philosopher Theodor Lipps explored what the Germans called *Langeweile* (literally, "long while").[12] For Waitz, boredom was about the flow of thoughts. As one thought begets another we generate expectations of where this thought train is headed. Boredom arises when those expectations are not met, thus there is a break in the flow of thoughts—the train is derailed.[13] Lipps suggested that boredom arises when we experience a conflict between our desire for "intense psychological activity" and an inability to be stimulated."[14]

Similar notions of boredom were explored by the forefathers of psychological research in the English-speaking world, the polymath Sir Francis Galton and the philosopher William James. Galton captures that notion of agitation that medieval monks called the noonday demon. Constantly finding ways to measure people and behavior, Galton wrote of audience members swaying to and fro and fidgeting during a dull scientific talk—clear signs of restlessness and boredom. In a speech given at the turn of the twentieth century, James bemoans "an irremediable flatness [that] is coming over the world."[15] For James, this flatness and its attendant boredom was driven by an increase in the *quantity* of information at the expense of *quality*.

What each of these early accounts of boredom hint at is the uncomfortable feeling of wanting to engage in satisfying activity but being unable to do so. Each of these accounts highlights a central tenet of boredom: the signal that we are mentally unoccupied.

An Existentialist Dilemma

James's "irremediable flatness" and Seneca's laments of the nausea born of monotony both point to a critical component of the experience of boredom—the sense that things lack meaning. In exploring the foreboding sense of angst brought on by acknowledging the absurdity of life, existentialist philosophers were among the first scholars to systematically address the role of meaning in boredom.[16]

For Arthur Schopenhauer, the pessimistic forebear of existentialism, the underlying reality of the world is most directly expressed through our bodily experience of desire. In other words, life *is* desiring, striving, and yearning. If life is a constant yearning, then the desires we harbor can never be satisfied in any lasting way; when one desire is sated, another rises, making desire itself all there is. Happiness—a momentary reprieve from desire—is always just about to happen. As soon as happiness does arrive, a new desire presents itself. According to Schopenhauer, then, we are predestined to suffer most of the time due to the ceaseless desire coursing within us. We are faced with two miserable choices: the pain of not yet fulfilling a desire or the boredom of not yet having a desire to pursue.[17]

Søren Kierkegaard, the Danish philosopher and progenitor of existentialism, also associated boredom with the struggle to find or make meaning. When we fail to adequately make meaning, we catch a glimpse of an impoverished and impotent self.[18] In *Either / Or,* Kierkegaard's hedonic raconteur states that "Boredom rests upon the nothingness that winds its way through existence; its giddiness, like that which comes from gazing down into an infinite abyss, is infinite."[19]

One understanding of Kierkegaard's message is that "boredom is a root of all evil" precisely because we seek to avoid it at all costs.[20] Diversions from boredom actually *increase* its stranglehold. Were we not so preoccupied by the need to escape from boredom it could point us toward another way of being, one where passionate commitment to a life purpose becomes our guide.[21] Indeed, the second half of *Either / Or* asserts that when we choose to move away from the hedonic and toward more ethical ways of being, boredom ceases to be so troubling.

One final existentialist necessary to consider in our quest to define boredom is Martin Heidegger.[22] Heidegger asks us to first imagine sitting in a train station, waiting for a train that is two hours late. Scanning the station provides only the shallowest of entertainment. We have a book or a phone, but they devour only brief periods of time before we are left wanting some new distraction from waiting. Heidegger refers to this as *superficial boredom,* directed at an external object or event that is not happening. In other words, time is dragging on.[23]

Next, Heidegger asks us to imagine being at a social gathering—some pleasant, innocuous affair—perhaps a work gathering celebrating a recent retirement. We discuss current affairs, trade stories of our children's latest achievements and foibles, and if in Canada, spend considerable time talking about the weather. It is not until later that we realize that all of that time, although pleasant enough, was utterly pointless! Perhaps we were engaged, but not in anything we would rate as meaningful. We sense our time was wasted. This is *boredom alongside* an activity not directly tied to one specific object or event, like waiting for a train. Yet, it is a third stage of boredom that is most important for Heidegger—*profound boredom.* This kind of boredom has no object or source. It is timeless and represents a kind of emptiness in which we get a terrifying view of reality.

So throughout the ages boredom has been linked to mundane routines (Seneca's "night overcoming day, day night"). And since no one thing can ever be counted on to fully satisfy us now or into the future, our daily struggles can feel as though they are empty of meaning. This is boredom's irony. On the one hand it highlights the inherent meaninglessness of existence while on the other it propels us forward in a never-ending search for something fresh and meaningful—something we *hope* will satisfy.[24]

Boredom on the Couch

Whereas the existentialists see boredom as a problem caused by a lack of meaning, psychoanalysts cast boredom as a solution to the problem of anxiety.[25]

According to classical psychoanalytic thought, our primal desires, buried deep under layers of socialization, are disturbing to us. Becoming aware of such desires threatens our sense of self, as well as the social order—we are afraid of our own desires. One coping strategy is to simply push unwanted desires out of our mind. In their wake, however, what remains is the strong feeling that we want to do something, combined with an inability to say precisely what that something is. We have banished the details of the desire to the dungeons of our subconscious. Confronted by a sense of wanting but without a specific target for our desire, we experience a restless tension and turn to the world in a futile attempt to find something compelling that will quell the wanting. Boredom, then, is the price we pay to stay emotionally safe.

In one of the early psychoanalytic accounts of boredom by Ralph Greenson, boredom is characterized as a restless, agitated state.[26] Greenson describes a patient for whom boredom arose from a need to keep depressing impulses at bay. In fact, for his patient the absence of boredom "led to either severe depressive reactions or to impulse ridden behaviour." He goes on to claim that for the bored person, "Tension and emptiness is felt as a kind of hunger—stimulus hunger. Since the individual does not know for what he is hungry, he now turns to the external world, with the hope that it will provide the missing aim and / or object."[27]

For the psychoanalysts, boredom represents avoidance of deeper psychological problems. But this leaves us stuck in other ways. Anything we can think to do won't satisfy because it is too far removed from the original desire.[28] Unaware of our emotions, we are adrift without direction.[29]

If existentialism highlights a paralysis due to meaninglessness, psychoanalysis emphasizes boredom's association with anxiety. Our attempts to cope give rise to absurd binds. The British psychoanalyst Adam Phillips writes of boredom as "that state of suspended anticipation in which things are started and nothing begins, the mood of diffuse restlessness which contains that most absurd and paradoxical wish, the wish for a desire."[30]

This is a paraphrasing of a line from Leo Tolstoy's novel *Anna Karenina*—"boredom—a desire for desires."[31] So, according to the psychoanalysts, boredom arises whenever we are threatened by what we really want.

Lack of life meaning and deep internal conflict; these seem to be uniquely human problems. Eric Fromm, a sociologist, psychoanalyst, psychologist, and philosopher of the twentieth century, famously declared, "Man is the only animal that can be bored."[32] Was Fromm wrong? Is boredom a uniquely human experience? Watching your cat chase after a laser pointer, it's hard to imagine she experiences existential angst or anxiety about unacceptable desires.

Beyond Human Boredom

We have little trouble believing that animals engage in play. Whether it is tiger or lion cubs wrestling, an elephant calf sledding down a muddy slope to crash into unsuspecting adults,[33] or killer whales tossing seals high in the air to bat them away with their tail flukes,[34] it seems plausible, even obvious, that animals play. Early theories on the function of play cast it as training for skills needed for adulthood or serving an important role in social bonding. Neither is the whole story. Young animals that play do not simply grow to become better hunters or have more friends. A more recent explanation of play behavior in animals describes it as a short-term advantage that would also work for humans—that is, play reduces stress.[35]

If we accept that animals play, and perhaps for some of the same reasons humans do, then shouldn't we also accept that animals have other experiences that we once thought were uniquely human? In other words, can animals be bored? If play helps animals cope with stress, then boredom might arise when the animal is prevented from engaging in behaviors it would normally choose—play or otherwise. It has long been recognized that there are detrimental consequences for animals reared in impoverished environments, from increased stress and poor coping strategies[36] to negative effects on brain development.[37] The converse is also true—environments rich in variety work to promote neural development. Importantly for our story, increased stress in animals has led some to suggest that those animals reared or housed in bland environs exhibit behaviors akin to boredom. Françoise Wemelsfelder, a scientist at Scotland's Rural College, has long been a champion of the idea that animals can indeed become bored.[38] For Wemelsfelder, it is the confined environs that captive animals are kept in that is the main culprit. Such confinement clearly restricts the available options for action for the animal. They are reduced to a restricted range of stereotypical behaviors that in some sense are pointless—they do not reflect the behaviors the animal would normally be capable of deploying in the wild. As we will argue throughout the book, when humans are bored, we too are confronted by a challenge to our agency—a feeling that our capacity to be the author of our own lives is being challenged or constrained in some way.

But how can we be sure that an animal is truly experiencing boredom and not something else? Rebecca Meagher and Georgia Mason from the University of Guelph conducted a study intended to discriminate between anhedonia, apathy, and boredom in captive bred Black mink.[39] Anhedonia—the inability to experience pleasure—is associated with depression in humans.[40] Apathy is considered distinct from boredom in that it reflects a lack of interest coupled with low motivation to redress the circumstance. In contrast, boredom is characterized by a strong drive to be doing something. In other words, the anhedonic person can't feel pleasure, the apathetic person doesn't care, and the bored person wants to be engaged. The problem, of course, is that we can't simply ask an animal to tell us when they're bored (or apathetic or anhedonic). But we can measure behaviors in response to the introduction of novel stimuli, which is precisely what Meagher and Mason did.

Two groups of animals were tested—mink housed in unenriched cages and a group housed in enriched cages that allowed for more varied, exploratory behaviors. Both groups were eventually shown objects classed as either aversive (the odor of a predator), rewarding (a moving toothbrush—the equivalent for minks of a laser pointer for cats), or ambiguous (a plastic bottle). The researchers then measured time to contact, duration, and amount of contact with the novel objects. The logic was that an apathetic animal should show decreased interest in all objects. In contrast, an animal exhibiting anhedonia should be less interested in only those objects considered rewarding. That is, the animal can't experience pleasure and so will not approach things normally seen to be enjoyable or positively rewarding. Things would be different for the bored animal. The researchers argued that a bored animal would engage indiscriminately with any and all objects. In other words, if the mink in unenriched cages were indeed bored, then any new object should satisfy their need to engage with the world. The researchers measured which type of objects the animals interacted with, as well as how quickly they went to the objects when they became available.

Mink in the unenriched cages were quicker to make contact with *all* object types—including the aversive odor of a predator! It seems the mink were not depressed or disinterested, they were desperate for stimulation, a clear sign of boredom. Consumptive behaviors—how many treats the animals ate—were also measured, and the mink housed in unenriched cages ate more treats than those housed in the enriched environments. Eating out of boredom is something humans do too. Even if we avoid the use of the word boredom altogether, what this study shows is that animals raised in bland environments become sensitized to new avenues for action.

Of course, all of this relates to animals in captivity. Do animals in the wild get bored? They likely do, but only for short periods of time. In unconstrained, natural environments animals are free to choose their next action. In captivity, their circumstances impose constraints that mean the animal is doomed to experience a monotonous life; they are caged and prevented from engaging in the full suite of behaviors they would normally exercise were they in the wild.

So for humans and animals alike, the key is that we must be self-determined and engage the world on our terms; we must be free to make choices based on what matters to us. Variety and excitement by themselves are not enough. In fact, excessive variety and constantly arousing events that are out of our control would not be comforting at all. This would feel more like anxiety or even mania. Choice is critical. To make a choice, one activity must be judged to be more important or satisfying than another. And research has shown that animals, like humans, tag different activities as more or less desirable—they make choices about how to express themselves and thus are subject to boredom.[41] So what boredom in captive animals reflects is the mismatch between the normal application of skills and capacities and the restricted possibilities for action offered by the current environment. If *where* we are prevents us from fully expressing *who* we are (or what we are capable of), boredom ensues for both human and animal.

Your Brain on Boredom

So our definition of boredom suggests it has a long history and is not unique to humans. But is boredom a fundamental fact of our biology? Research trying to pin down the biological signature of boredom attacks the problem from many angles, looking at changes in physiological measures like heart rate, measuring electrical brain signals and exploring the network of brain activity in action using functional brain scans. All of this work is in its infancy. Researchers have only just started to peer into our minds to better understand the neural signature of boredom.[42]

Our own attempt to more closely zero in on the brain activity associated with being bored involved having people watch one of two video clips while we scanned their brains in an MRI machine. One was intended to make them bored—eight minutes of two men hanging laundry. The other was a clip from the *Blue Planet* series by the British Broadcasting Corporation (BBC)—anything but boring. The results were fascinating. Boredom was differentiated not by how much activity was evident in the brain but by how two different networks were linked. One part, the insular cortex, has among other things the job of signaling that something important is present in the environment, helping us devote our mental faculties to that information.[43] The default mode network, on the other hand, is active when there is nothing worth paying attention to in the external environment. It's also active when we turn our focus inward on our thoughts and feelings. When bored, these two parts of the brain—the insular cortex and default mode network—were anticorrelated, meaning that as one area became more active, the other became less active. While more research is needed, we believe this pattern may reflect continued failed attempts to engage with what is a monotonous, uninteresting series of events. Critically, these two parts of the brain were not linked in the same manner when people were just resting or watching the clearly more interesting BBC clip. All of this suggests an intriguing idea: the bored brain is not simply a brain with nothing to do but a brain that is hoping for, anticipating the *possibility* of something to do. At this point the research is so preliminary that our conclusions are little more than informed speculation, but the existing findings are intriguing and, at the very least, suggest boredom is associated with a distinct neurological state that is rooted in our physiology.

A Modern Definition for an Ancient Experience

Boredom has been with us a long time, long before Dickens's eloquent treatment of it. To be bored is to be painfully stuck in the here and now, bereft of any capacity for self-determination, yet driven to find something that we can engage with. Over the past few decades we have seen a surge of interest in defining boredom. Wijnand van Tilburg and Eric Igou suggest boredom is best defined as a lack of meaning, whereas Erin Westgate and Timothy Wilson suggest boredom is the result of both a lack of meaning and deficient attention. Thomas Goetz and colleagues suggest that in educational settings there are as many as four or five different kinds of boredom. Andreas Elpidorou sees boredom in a positive light, acting as "a push" to engage in something different. Heather Lench and Shane Bench support a similar account in which boredom arises as our emotional responses to a situation diminish in intensity, galvanizing us into action.[44]

But what do *we* mean by boredom? To tackle this directly we examined how experts from various disciplines and everyday people define the experience of boredom. We found remarkable agreement across all groups—*boredom is the uncomfortable feeling of wanting, but being unable to, engage in satisfying activity.*[45]

Boredom is the feeling we get when we want to engage our mental capacity but fail to do so, leaving our mind unoccupied. It's well known that we feel our emotions, but we don't often stop to consider that we also feel our thinking. And that is what we are talking about here. It's not about *what* we are thinking but rather *how* we are thinking.[46] Consider the positive feeling associated with easily understanding something we are reading, or the feeling of being on the cusp of solving a puzzle, or the feeling of being so absorbed in something that we are blissfully unaware of time passing. In contrast, when we have to exert mental effort it can feel downright miserable. We are not the first to define boredom as a feeling of thinking; the idea has roots in Theodor Waitz's notion that boredom arises from a break in the flow of thought. More recently, Jean Hamilton and colleagues proposed that boredom arises from "the cognitive, information-processing act in attention."[47]

Note that in saying boredom is a feeling associated with our thought processes we are deliberately not classifying it as an emotion. Emotions, unlike feelings, are generally thought to possess multiple components. Emotions are triggered by specific eliciting events that are relevant to basic needs and serve to organize us into a response specific to the triggering event. In contrast, boredom as a feeling of thinking is not linked to an external event in the same manner but is a felt experience of an ongoing cognitive process.[48]

What, then, is happening in our mind when we feel bored? We argue there are two underlying mechanisms at play. First, boredom occurs when we are caught in a *desire conundrum,* wanting to do something but not wanting to do anything. We can't muster up a desire to do anything in the moment. We may wish for a different set of circumstances, and in that sense we are frustrated, but critically we can't cultivate a desire for anything that is currently doable, and we are bored. Second, boredom occurs when our mental capacities, our skills and talents, lay idle—when we are mentally unoccupied. When both mechanisms occur, we are bored, and without them, we aren't. Each mechanism positively reinforces the other. To be clear, these mechanisms are not *causes* of boredom, rather they *are* boredom. Indeed, just as we have deliberately chosen not to define boredom as an emotion, we have also deliberately chosen not to define boredom in terms of its cause.[49]

Boredom is what it feels like to have an unengaged mind. What does it mean to be unengaged? When we say a person is unengaged, we mean their mental capacity is under-utilized. Two factors determine how much of our mental capacity is being used at any given moment—how much capacity the activity demands and how much capacity we choose to devote to the task at hand. Some activities are more engaging than others; some simply don't require enough to fully occupy our minds. Imagine trying to remember the number 3 for the next thirty minutes—not very challenging. In fact, this is so easy you should have lots of unused mental capacity left over. And if you don't find something to do with this unused mental capacity, you will likely become bored. On the other hand, if we don't want to put our mind to what is going on around us, no matter how complex the opportunities, our mind will remain unoccupied. It is possible, however, to become engaged by our own internal thoughts when the activity at hand is too simple or when we can't bring ourselves to focus on what is in front of us in this moment.[50]

Evolutionary forces have shaped us to experience the discomfort of boredom when our cognitive resources are underutilized. Just as we feel hungry when our body is malnourished, we feel emotional discomfort when our mind is undernourished, and we are motivated to remedy the situation. We are biologically predisposed to seek mental engagement, and boredom is the signal that keeps us searching for that engagement. This drive was more uniformly adaptive before the invention of such easy engagement machines as smartphones. Now with so many ever-present, tempting, quick, and easy outlets to occupy our minds, our drive to avoid the distress of being bored can lead us to some dark places.

The claim that we are hardwired to seek out mental engagement should not be confused with a need to exert mental effort. In fact, it seems the opposite is true—all things being equal, we seek activities that easily engage our minds. We are "cognitive misers."[51] Occupying our mind doesn't necessarily require exerting effort. We can relax and let our mind drift without any intention to think about anything in particular and still be mentally engaged. Wondering how we'd spend our lottery winnings and idly planning our future as prime minister or president are two good examples of the sort of fantasizing that occupies our minds with little effort.

In addition to a feeling of discomfort, there are four telltale signs that boredom is upon us. These signs are not strictly part of the definition, but they bear mention as cardinal elements of the experience. The first sign we may notice is that time drags on when we're bored. The German word for boredom, *Langeweile,* pretty much sums up the experience. Without something to occupy our minds, we turn to the only thing left—tracking the mere passage of time. Think of those times you find yourself waiting for your number to be called so you can get your driver's license renewed. You sit in an uncomfortable chair with a gaggle of strangers sharing the same fate, the air-conditioning malfunctioning and nothing to do but watch the LED display tick over. Time drags on. A second sign of boredom is the struggle to concentrate. In the face of monotony and a lack of desirable options, our attention wanes, our minds wander and boredom rises. Imagine those interminable meetings you've had to suffer through, knowing you ought to be paying attention to what this month's numbers are saying, or what upper management sees as the cause of a 1 percent rise in days lost to illness, but you just can't muster the effort to give it your full attention. Third, when bored, whatever it is we are doing feels pointless. Simply put, activities that do not occupy our minds, or that we can't bring ourselves to want to do, hold little value. Not only is it hard to pay attention during that interminable meeting about a miniscule rise in sick days, the whole thing feels like a colossal waste of time—time you will never get back. Fourth, boredom is associated with a combination of lethargy and restlessness. When we are bored, our energy levels can be all over the place. One moment you find yourself lying listlessly on the sofa, unable to will even the tiniest of movements. The next moment you're all keyed up, urgently pacing around the living room looking for something to do. Nothing can satisfy that restless itch. Each time your eye lands on something, you think to yourself, "Nah, not worth it."[52]

It is important to highlight that what we have described is a definition of the actual, in the moment feeling of boredom, what psychologists and social scientists refer to as *state boredom.* Throughout the book, we will simply use the term *boredom* when we're talking about this state experience. For some, the feeling of boredom comes around more frequently and is felt more intensely. This reflects a personality-based predisposition to the experience that psychologists and social scientists refer to as *trait boredom proneness.* In the next chapter we explore the ways in which boredom can be caused by both situational factors and personality factors. However, regardless of the cause, the feeling of boredom and its underlying mechanisms do not change. In other words, there are not multiple types of boredom. We argue for a singular definition of the experience. In our view, defining different kinds of boredom only serves to muddy the waters.[53]

Pinning down a definition is something scientists are compelled to do, but in the real world this is a messy affair. The thing Lady Dedlock first called boredom has been tormenting us under different names for some time. Nevertheless, with our tentative definition in hand we can now turn to unearthing its cause.

Boredom can strike virtually anywhere—but some places are particularly dangerous. Nobody is completely untouched—but some suffer more than others. We next examine both external situations and internal personal factors that provide insights into the origins of this elusive state of mind.

CHAPTER 2

A GOLDILOCKS WORLD

• • •

You started the day excited. It's your first trip out west, and you're looking forward to taking in the majesty of the Rocky Mountains, the Columbia Icefields, the wildlife, all of it. It's been on your bucket list since high school. But it seems the regulations and rigmarole attendant on a day of air travel are actively out to get you, to dampen your spirit and bore you out of your skull.

Checking in to your Air Canada flight was supposed to be simple. Most of the grunt work was completed online, so you imagined a kind of fast-lane experience. It wasn't to be. Some insignificant piece of missing information meant that you found yourself at the check-in booth waiting for the issue to be resolved. Half an hour dripped by. Then an hour. Keys on a keyboard were furiously tapped, phone calls to superiors were made, and finally, the boarding pass was handed over. Foolish to think this was the end of it.

At security you looked for the fastest lane. Pointless, really, and you knew it. You've done the same thing before at the grocery store and inevitably have found yourself stuck behind the grandma intent on paying the bill in dimes and nickels painfully extracted, one at a time, from the world's smallest change purse. But you had to try anyway. The far right lane seemed to be moving well, so you shuffled around to position yourself there.

You didn't look at who was in front of you. A family with three carry-on bags apiece; each big enough to hold your entire luggage. And the security agents decided each one needed to be searched. Small jars of makeup, shampoo and conditioner in full-size bottles, sippy cups, and a pile of juice boxes and water bottles. What rock have they been living under? All of it inspected, argued over (of course they were indignant), and ultimately discarded. Your fast lane was now glacial.

And it wasn't over once you got past security. You had planned to blissfully sink into your John le Carré novel tucked away in your carry-on before boarding, but somehow you just weren't feeling it anymore. You could have picked up a magazine across from the departure lounge or done a bit of duty-free shopping. Nope. Convinced none of that was going to work, you sat immobilized. There was now nothing that you wanted to do—yet you most definitely wanted to be doing something! In the end, all you could do was wait, forced to attend to the only thing left that mattered—the slow march of time. Boredom washed over you in restless wave after restless wave.

• • •

We tell a bored person to just snap out of it. Or we avail them of options we assume ought to work. Read a book. Go for a run. Watch television. Call a friend. Essentially, we are telling them what they already know—there are a multitude of possibilities out there for them to engage in. Our immediate response to boredom in others highlights our ignorance of the underlying mechanisms and the originating causes of boredom. The bored person knows all too well that there are *options* for engagement. They just can't *engage* with any of those options. If they could, they would. If boredom is a failure to launch into an action, then telling someone to do so does not magically make it possible. We would not tell someone who is drowning and unable to swim to simply swim to shore.

Figuring out the causes of boredom is a tricky thing. In short, it's complicated. As we've already seen, there are two mechanisms in play when we feel bored—underutilized cognitive potential and a desire conundrum. Some causes of boredom directly elicit one underlying mechanism, whereas others give rise to the other. Our purpose in this chapter is to follow the thread further back to understand the various causes of boredom. We begin our search for boredom's causes by elaborating the mechanisms that underpin the feeling of boredom.

Boredom is both the pain associated with an unoccupied mind and also the pain associated with what we're calling a desire conundrum. The bored person is burdened by the uncomfortable feeling of being mentally unengaged. Whatever it is that they are doing, thinking, feeling, or imagining is not sufficiently using their cognitive resources. At the same time, the bored person wants to do something but can't get started on anything—a kind of failure to launch. This creates an intense bind—a desire to do something but a failure to attach that desire to anything that can be done *right at this moment.* Boredom, then, is undefined, directionless wanting—Tolstoy's "desire for desires." In some ways, boredom resembles the "tip-of-the-tongue" phenomenon, that moment when you can almost but not quite remember the name of that guy who starred in that movie.[1] "Was it Mel Gibson? No, that's not right . . . Bruce Willis? No, not him either. Keanu Reeves, that's it!" That feeling consists of twin forces: the absence of something and the drive to fill in the absence. At least in the tip-of-the-tongue phenomenon you have the sense that the name is tantalizingly close, and you'll get it in the end. With boredom there is no such optimism. Instead, there is a frantic casting about to find something to fill the void and relieve the pressure. Somewhat akin to trying on different jackets until we find the one that fits, the bored person tries to work backward from the feeling of relief to discover a workable desire. When bored, we ask too much of the world. We expect the world (or others in it, as in the case of children begging their parents to *solve* their boredom) to clarify what it is we want to do. An extended episode of boredom arises when even then we fail to find a target for our skills and talents.

Situations sometimes constrain us so that our desires are blocked. Sometimes we can't do what we want to do, such as go sailing when we have to work, or else we have to do something we don't want to do, such as doing our taxes. Both circumstances characterize frustration, not boredom. Being frustrated or prevented from engaging with the world in the way you want to can lead to boredom eventually, but they don't *necessarily* cause boredom. When we find ourselves in circumstances that constrain our options for engagement, we might cast around for other options and still discover that we don't want to do anything that is doable in the moment. And when this happens, one of the two key mechanisms of boredom is ignited—the desire conundrum—and the second follows in quick succession—an unoccupied mind. With both in play we become bored. In these situations we might say that we are both frustrated about not being able to do what we wanted to do in the first place, but also bored because we can't bring ourselves to want to do anything that is on offer. On other occasions, perhaps grudgingly, we become engaged with the options at hand. Such options might include continuing to do the undesirable task (getting the taxes finished) or seeking escape through daydreaming (looking out your office window and imagining sailing). Successfully continuing on the task or letting our minds wander and daydream means that, while we might still say we're frustrated, we are at least not bored.

So boredom is not frustration. Frustration occurs when we're prevented from achieving a clearly stated goal: we want to go sailing, but we have to work instead. Boredom occurs when the most pressing goal we have is to have a goal, and yet none presents itself. The bored person is tormented by a wanting without knowing the conditions for satisfying that want.[2]

Schopenhauer described this situation precisely: "We consider ourselves fairly fortunate if there is still something to wish for, and to strive after, to keep up the game whereby desire constantly passes into satisfaction, and satisfaction into new desire—if the pace of this is swift, it is called happiness, and if it is slow, sorrow—and does not falter and come to the standstill that shows in dreadful, stultifying boredom, in lifeless yearning without a definite object, a deadening languor."[3]

Boredom presents as a thorny dilemma. We want to do something, we have a desire to be engaged, but nothing we see on the horizon seems like a viable option to satisfy. This is what we are calling the boredom conundrum.

We could try to shut down desire itself by seeking refuge in sleep, succumbing to apathy, or cultivating the practice of quietism. If the pressure to do something abates, we will no longer be bored but content with *nondoing.* Alternatively, we could force ourselves to do something until it holds our attention and satiates our wanting: we could dive down the rabbit hole of the Internet, pick up Candy Crush for the nth time today, or try yet again to read that challenging classic, *War and Peace.*

A third path, discovering an actionable desire, might be possible, but it is not something we can make happen. Rather, it happens to us. It is like trying hard to fall asleep—the harder we try, the more intractable the problem feels. We can, however, create the conditions that are likely to foster sleep. Likewise, we can establish and nurture the conditions that are likely to help us find what it is that we want to do. Sometimes it amounts to not trying so hard, and like the tip-of-tongue phenomenon, we find our desire when we back off.[4] Perhaps we would do well to heed the advice of Hermann Hesse's Siddhartha: "What could I say to you that would be of value, except that perhaps you seek too much, that as a result of your seeking you cannot find."[5]

So our mind is unoccupied and we want to do something, but we can't figure out what we want to do—that's boredom in a nutshell. But how did we get here? What causes boredom? Endless political debates, listening to someone recount the same story you've heard a thousand times before, sitting through a tedious meeting at work—the list of *boring situations* is legion. Indeed, external factors are an important cause of boredom. In fact, there are four prominent external factors that lead to boredom: monotony, lack of purpose, constraint, and poor fit between our skills and the challenge of the moment.

The Four Horsemen of Boredom

If you stare at the small dot in the black and white Canadian flag in Figure 2.1 for thirty seconds and then look at the dot in the blank space underneath, you'll learn something important about your own brain—it doesn't like staring at the same thing for a long time. The image you see—the Canadian flag in reversed contrast—is called an adaptation after-effect. What it tells us is that even our basic sensory systems need change and variety to function optimally.[6]

Monotony, the opposite of change and variety, was one of the first of the horsemen—that is, external causes—to be explored by researchers, starting in the workplace. In the years

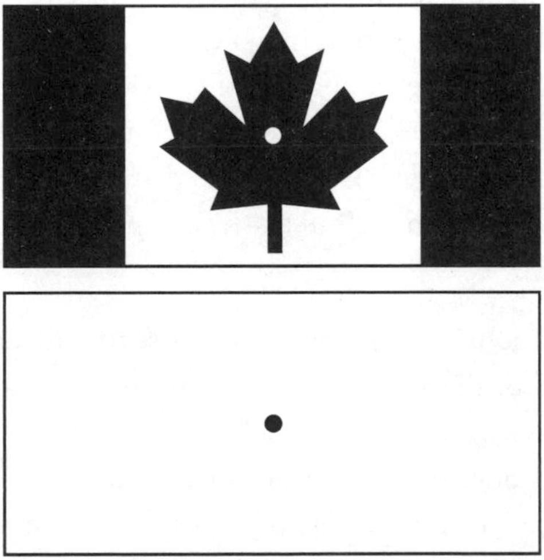

Figure 2.1. An adaptation illusion of the Canadian flag. First, stare at the small white circle in the image of the Canadian flag for thirty seconds. The illusion is stronger the longer you stare and the better you are at not looking away from the dot. Then look at the small black dot in the blank space beneath. You should see a reverse image of the one you just stared at.

between World War I and World War II, a shift occurred in the workforce, away from physically demanding and dangerous work toward lighter, repetitive, and mechanized factory work. Workers were increasingly being asked to do the same simple task over and over again for long periods.[7]

J. C. Bridge, the senior medical inspector of factories in England, described the plight of the worker in 1931:

> It is true that the pleasure of the craftsman is being crushed by the steady increase in mechanized processes, the result of which is seen in the tendency to rise of sickness rates for "nervous disabilities." . . . Repetition processes undoubtedly create a weariness not expressed in physical terms but in a desire by the worker for temporary relief from the enforced boredom of occupation in which the mind is left partially or entirely unoccupied. . . . More interest in processes that are themselves dull must be created. Selection of workers is in this problem only of limited value; there are more dull tasks than people suitable for them.[8]

Monotonous tasks are boring because they demand our attention but at the same time fail to fully occupy our mental resources. Monitoring an assembly line for the odd defective component does not use all of our mental capacity. But the moment we slip into a daydream, or our mind wanders to thoughts of what we might be doing that evening with friends and family, is the moment those defective components sneak past us.[9] That's the rub. Boring, monotonous tasks require *just enough* of our mental capacity that we can't complete them mindlessly, but not enough of our mental capacity to satisfy our need to be engaged.

Hugo Münsterberg, a pioneer of industrial and organizational psychology, was perhaps one of the first to identify and study the problem of monotony in the workplace. For Münsterberg, workplace monotony was the "subjective dislike of uniformity and lack of change," which he observed was not an objective quality of the environment but the result of individual judgments. One man's monotony is another man's variety.[10] He described a remarkable man who had made the same 34,000 uniform movements daily. Every workday for fourteen years he slowly pushed a metal strip into an automatic drill at the right pace and the right angle so that holes could be cut into it at a precise place on the strip. Münsterberg spoke with this man because, from an outsider's perspective, he thought this must be an excruciatingly boring job. To his surprise, the man indicated he was not bored at all. In fact, he found the work interesting and stimulating. He even claimed to enjoy the job more as the years passed. Münsterberg was astonished:

> I imagined that this meant that . . . the complex movement had slowly become automatic, allowing him to perform it like a reflex movement and to turn his thoughts to other things. But he explained . . . that he still feels obliged to devote his thoughts entirely to the work at hand, [in order to maximize his wages which depended on the number of holes made but] . . . that it is not only the wage which satisfies him, but that he takes decided pleasure in the activity itself.[11]

The monotonous activity had both intrinsic and extrinsic value to this worker, which made all the difference to him. Not everyone finds purpose in repetitive behavior, however, which leads us to the second of the four horsemen of boredom—purposeless activity. Repetitive behavior by itself can't cause boredom; it must also lack any perceived value. In this case the worker could become fully occupied by a task that, on the surface, seemed to offer little to sink his mental teeth into. The more he attended to his task, the more engaged he became. When we pay attention, we notice the little things. When we notice things, we want to look again. Then, before you know it, we are fully engaged experts in nuances others can't fathom or find simply uninteresting. That's the power of motivated attention to obliterate boredom. Without that motivation, monotony is a powerful driver of boredom.

So we can transform a monotonous situation into a valuable one if we have reason to do so. For Münsterberg's factory worker it was the powerful driver of money and providing for his family. More recent studies indicate it can be as simple as convincing people that the task they are doing is good for their health or will lead to improved performance on tests and enhanced job prospects later in life.[12] The power of a good reason for doing something works by making us want to do something we otherwise would not want to do. But what if we are forced to do something and can't find any reason for doing it? This question brings us to the third horseman of boredom, constraint. Being forced to do something or being prevented from doing something can certainly lead to boredom.

As early as 1937, Joseph Barmack at the College of the City of New York noted the link between constraint and boredom.

For him, boredom was a state of "super-satiation," or the feeling that occurs when a person is compelled to keep doing something beyond the point at which they want to stop.[13] Indeed, the early work on monotony was about boredom under workplace constraints.

Monotony and constraint feed off each other.[14] As monotony drags on, our energy levels dip—a kind of habituation to the repetitive, unchanging nature of what we're doing. Think of things like data entry—hours spent entering numbers on a spreadsheet. Maybe you're entering the hours your employees worked this month or tracking inventory changes, but it's not enough to hold your interest. Arousal levels can dip so low that we start to get sleepy. In some circumstances this would be no problem—just take a short nap or surf the Internet. But if the thing we are doing can't be ignored, we must find a way to bump up our energy levels or mistakes will happen.[15] Not all professions lend themselves to naps (think air traffic control). Fidgeting is sometimes seen as serving that purpose of bumping up our arousal.[16] But what is good for us is not necessarily good for the task at hand, and it is no simple matter to get sufficiently energized when the task is repetitious and lulls us toward sleep.

This battle between diminishing energy levels and the continued need to pay attention may be phasic, ebbing and flowing as monotony keeps getting the upper hand, demanding that we exert more effort to keep sleep at bay. Boredom is a condition of fluctuating levels of arousal.[17] But we would not experience this fluctuation if the situation didn't compel us to keep doing the monotonous task. Without constraint, there would be no boredom. You could simply choose sleep or opt to do something else. Either way, boredom would be eliminated.

Mark Scerbo and colleagues from Old Dominion University in Norfolk tested the idea that constraint leads to boredom directly. In their study, one group of people could end a monotonous task whenever they wanted, and another group had to keep going until the researcher ended it. The mere freedom to stop resulted in lower reports of boredom even though both groups spent exactly the same amount of time on the dull task.[18]

Simple, repetitive, unchanging tasks that we don't care about yet are forced to do—that has been the focus of most research on boring situations. Under-challenging situations that we can't walk away from do indeed lead to boredom. But that's not all.

As Odysseus knew all too well, navigating a single hazard is relatively easy, but place yourself between two monsters and things get impossibly tough. Ill-fated Odysseus had to sail the strait of Messina while avoiding Scylla (shoal) on one side and Charybdis (whirlpool) on the other.[19] Similarly, in order to avoid succumbing to boredom, we face the difficult task of navigating not just the peril of being under-challenged but also the danger of being over-challenged. To avoid boredom we need to find a fit between our skills and interests and what is possible—a kind of Goldilocks zone. Failure to find that zone is the fourth horseman of boredom.

Video games provide a good analogy. Imagine playing Tetris at the same beginner level for an hour. If you are unable to level up, the game becomes repetitive, your skills are underutilized, and you become deeply bored. Now imagine starting a game of Tetris at the most challenging level any human has ever played. You are rapidly overwhelmed. You needed to work up to it, pushing the ceiling of your skills. In the first case, all we have is redundancy—nothing new. In the second, all we have is chaos and noise—nothing we can make sense of.[20] In both cases we can't become engaged, and boredom quickly sets in.

While it may seem obvious in light of our video game analogy that optimal levels of challenge and fit between our skills and the task at hand are critical for warding off boredom, there has not been a wealth of scientific research to directly test the idea. In one study, we had two groups of people play the children's game of rock, paper, scissors against a computer opponent. One group artificially won all the time, and the other group always lost. Each group rated how much they felt they were in control of the task, and surprisingly those ratings spanned the full range of possible responses. Some people who always won did not feel fully in control, and some who always lost still felt as though they had some say in how the task panned out. When we then looked at how bored they were, it was the people who felt completely in control or those who felt they were totally at the mercy of the computer opponent who were most bored. Recast as a consequence of challenge, these data suggest that when there is no challenge (always winning) or too much challenge (always losing), we feel most bored.[21]

In another study, we had two groups of people watch different twenty-minute videos. In one, a ridiculous mime taught extremely basic English vocabulary in an excruciatingly slow and repetitive manner. In the other, a brainiac taught advanced computer graphics using incredibly complex mathematics and impenetrable charts. In both cases, we told people they had to pay attention because they would have to answer questions about the video at the end; in both cases, people could barely endure the ordeal. Boredom levels didn't differ. Being under-challenged by a boring mime or over-challenged by complex math brought people to exactly the same miserable place.[22]

The four horsemen of boredom, representing risk factors in our environment, must be avoided, it's true. When bored, we are quick to blame the world for letting us down. But that is not the whole story.

The Causes Within Us

We're not innocent bystanders. What we bring to the table matters. Research shows that some of us are more prone to boredom than others. How we react and the skill we bring to the situation determines whether or not we will be bored.[23] As we see it, there are five main internal causes of boredom: *emotion*—our felt sense of how we are in the moment; *biology*—our ability to be alert and responsive to the environment; *cognition*—our ability to focus and think about the world around us; *motivation*—the push to engage in something; and *volition or self-control*—the ability to establish and follow through on a plan. Weaknesses in any of these internal domains can put us at risk for boredom.[24]

Emotions reveal our relationship to what is happening around us. In other words, emotions tag things as important in various ways. Without them we would be adrift, unable to identify what matters. Without emotions to signal the value of things, our world flattens and loses its color. There would be little reason to do one thing over another and no reason to do anything in particular.

It should come as no surprise then that poor emotional awareness has been linked to boredom.[25] Without the ability to label how we feel or what matters to us, it becomes difficult (if not impossible) to identify a plan of action. There are a variety of explanations as to why the bored person lacks emotional awareness, but they largely converge on the notion that the bored person fears emotions and attempts to avoid them.[26] One existentialist perspective proposes that emotional numbness and its attendant boredom may be a response to a fundamental background anxiety we feel when we realize that we are the authors of our own lives, a realization that can be profoundly unnerving.[27] It can feel a whole lot easier to find some external rule to live by or to blame external forces for our state of affairs than to take our own lives in hand. In this sense, boredom is the price we pay to stay insulated, cut off from threatening emotions and realizations.

But is there research evidence to support this model of poor emotional awareness and boredom? Steven Hayes and colleagues at the University of Nevada developed a scale to measure what they called "experiential avoidance," or the tendency to avoid or escape unwanted feelings. People high in experiential avoidance find their emotions threatening and try to avoid them.[28] Research from our lab has shown that people who avoid their feelings also report feeling bored more often. Furthermore, an inability to accurately label emotions is prominent in people prone to frequent bouts of boredom.[29] Avoiding our emotions and lacking emotional awareness effectively robs us of the enlivening and animating role that emotions can play in our lives. Without rich, differentiated emotions, we experience a world drained of significance, making it difficult to identify what is meaningful to us. (We turn to the role of meaning in more depth in Chapter 7.) For now, we suggest that avoiding emotions, lacking emotional awareness, and failing to find meaning lead to boredom by making us incapable of identifying a valued activity. However, even if we find something to do, we won't be able to effectively engage unless our brain is sufficiently responsive and energized.

There is a long history behind the idea that an internal struggle to keep alert and responsive to the world causes boredom. For example, Münsterberg devised a series of experiments in 1913 to isolate the particular characteristics (within the eye of the beholder) that lead to feelings of monotony. He concluded that some people couldn't observe the same discrete event repeated without blurring them together. Each time they saw the event, they were less likely to register it the next time. People who have this quality are most distressed by repetitive tasks that require attention to the same thing over and over again (assembly-line work or data entry are good examples). These tasks are more difficult for them. Time drags on as each moment becomes indistinguishable from the next. Everything new is old again.

In 2009 Yang Jiang and colleagues demonstrated that people who are prone to boredom have slower and less pronounced neurological responses in the front part of the brain when they are repeatedly shown the same visual image. Boredom-prone people, it seems, quickly become accustomed to things in the environment. What is fresh and new for the rest of us rapidly becomes the same old, same old for the highly boredom prone. Essentially, they need more novelty to pay attention.

So part of the problem for the highly boredom prone is that they quickly become neurologically unresponsive to things around them. Alongside this challenge is another problem faced by the boredom prone—the feeling that they are chronically under-energized. As a remedy for under-stimulation, the boredom prone rush around here and there looking for excitement to give them an energy boost. This is a critical component of the boredom experience: without enough energy to successfully engage our cognitive resources, we inevitably become bored. Those who have a lower default setting for arousal will lack the activation needed to engage successfully. And if they can't get that extra boost from external excitement, they will not be activated enough to become mentally engaged.[30] Being responsive to what is happening around us and sufficiently energized is not enough for successful mental engagement, however. We must also have adequate cognitive abilities.

What do we mean by cognitive abilities? Cognitive ability is, admittedly, a nebulous and broad term. Here we are primarily focusing on the ability to concentrate and control attention. That is, the ability to control what we pay attention to, tune out distractions, inhibit impulses, hold information in mind while thinking about it, and flexibly shift between thinking about different things. Controlling our attention is a fundamental cognitive skill that acts as the gatekeeper for what we think about and allows us to connect with our surroundings and inner sensations. People with weak attention skills will sometimes find situations too challenging and be unable to engage their mind with what is happening in the moment. Thus, if the definition of boredom is the unfulfilled desire to be mentally engaged, then weak attention skills are a logical cause of boredom.

People with neurological conditions known to impair attention often feel bored. A recent review found that boredom is a common and debilitating problem for people who have experienced brain damage.[31] Although the increased boredom seen in brain-injured people could occur for a number of reasons, such as tedious rehabilitation regimens and diminished opportunities for engagement, it is also likely that brain-injured patients experience more boredom because of diminished attention that is a direct consequence of their injury.[32] People with neurological conditions such as attention deficit / hyperactivity disorder or schizophrenia, report often being bored.[33]

The link between attention and the tendency to be bored is also found in people without neurological conditions. For example, people who report having problems concentrating also report feeling bored more often.[34] Chronic problems with attention and concentration may be a cause of chronic boredom; that is, they do not just co-occur, rather the attention problems seem to be the reason for the boredom.[35]

Those who struggle to deploy their attention effectively may find their challenges with boredom compounded. On the one hand, whatever it is they are doing will be less likely to hold their attention and occupy their mind. On the other hand, they won't be able to find as many things that they want to do because they know most things fail to hold their attention. Simply put, they need things in the environment to strongly capture and hold their attention because they struggle to do that for themselves. So while some need only a good book (and time to read it) to occupy their minds, those of us prone to boredom need an action movie complete with thrills and constantly changing events. Thus, impoverished attention may increase boredom by both reducing the ability to become engaged and diminishing the capacity to even name something they *want to do.* This second possibility leads us to another cause of boredom that resides within us—motivation.

It has been said that there are two kinds of people in the world; those who are motivated to maximize pleasure and those who are motivated to minimize pain. That is, some of us are motivated to search for the next pleasurable activity while others adopt a cautious, plodding path to avoid even the slightest of problems. Living at the extremes of either of these motivations can put you at risk for boredom.

A constant desire to maximize pleasure is unreasonable. At some point those who seek ceaseless pleasure will find the world underwhelming. Things take time, mundane chores are always waiting, and tedious meetings must be attended. It's OK to want excitement and pleasure, but when that motivation becomes extreme, we will be prone to boredom simply because the world is not constantly rewarding. An extreme motivation for pleasure, excitement, and variety makes it difficult to want to do what is possible in the moment because most available options are coded as not being nearly rewarding enough.

But what about the other extreme—the desire to minimize pain? Characterized as "playing it safe" and associated with patterns of avoidance behavior, this motivational approach to life might avoid pain, but it also restricts options for engagement. Boredom, we claim, arises when we are not mentally occupied. Even the person fearful of every shadow has this desire to be mentally engaged. Clearly, it is not possible to satisfy both the desire to avoid engaging with the world to minimize pain and the desire to engage with the world to ward off boredom—a mismatch of a different kind than we described for the pleasure seeker.[36]

Research in our lab confirms the notion that these two distinct motivational paths to boredom exist.[37] The two paths reveal themselves via the use of distinct scales purporting to measure boredom propensity. Choose the Boredom Proneness Scale (perhaps the scale most commonly used by scientists interested in boredom) and you find that it is those who seek to minimize pain who experience boredom the most. Use the Boredom Susceptibility Scale (actually a subscale of a larger tool used to examine sensation seeking) and you find that it is those who seek excitement and pleasure who are more likely to experience boredom. So, in terms of motivational styles, the two routes will get you to the same place: at risk for boredom.

Most of us are a complex mixture of motivational approaches. We might slave away at a difficult job because the paycheck at the end of the week is worth it. We might also pick up our dirty laundry from the floor to avoid angering our partner. But sometimes we are motivated to do something for the pure joy of it, even if it doesn't trigger a positive or negative outcome. Climbing a mountain simply for the challenge or learning a musical instrument despite having no ambitions for rock superstardom are both examples of intrinsic motivation—the activity is its own reward. In contrast, our decisions are sometimes shaped by extrinsic motivations such as the desire for a paycheck or fear of annoying a partner. When we are intrinsically motivated to do something, we do it for its own sake. We are exercising and developing our skills, fulfilling our need to be competent and self-directed.[38]

It is hard to imagine someone who is intrinsically driven to engage with the world being bored. You don't often hear "I really love playing guitar but I'm bored while I play!" Unfortunately, we don't know much about the individual personality traits linked to intrinsic motivation. The vast majority of researchers have studied it as a state—not a trait.

We do know that people who are motivated by the intrinsic rewards of competence and self-determination are less likely to experience boredom.[39] They see activities as opportunities to flex their cognitive and creative muscles rather than occasions to achieve any specific outcome: the doing is what matters most. We also know that if our needs for autonomy and competence are not met—when we fail to be the authors of our own lives—we are more prone to boredom.[40]

But being in control is not easy. Sometimes we have impulses and desires that are out of step with what others expect of us. Other times we may even find ourselves behaving in ways that are inconsistent with what we want, as strange as that must sound. Self-control is essential to living harmoniously with others and critical to achieving our personal goals. The capacity for self-control ranges from being able to sit still and avoid fidgeting during a job interview to planning and deftly executing a multiyear course of study to prepare for a new career. As we have seen, motivation can point us in opposite directions, either avoiding problems or achieving good things. Self-control is about taking our priorities and turning them into actionable goals that we successfully pursue.

Given the important function of self-control, it is perhaps not surprising that people who report feeling bored often also report difficulty with self-control.[41] This relation is robust. Even after ruling out the impact of things like age (getting older and wiser is in part about gaining more self-control) and sex (males are consistently more likely to be prone to boredom than females), the link between the tendency to feel bored and lower levels of self-control remains strong.[42]

"Self-control" is a broad catch-all term that covers a wide range of important abilities. One such ability, which we call "self-direction," refers to a capacity to exhibit self-control in service of something we want to do. This includes figuring out *what* we want to do in the first place and then regulating our thoughts, feelings, and actions in order to accomplish whatever that is. Self-direction is a distinct type of self-control. Others, such as impulse control (not reaching for the third doughnut at the morning meeting) and inhibition (avoiding those fidgety movements in an interview) have less to do with formulating and implementing a plan for action. Self-direction is about giving expression to our desires. We might be motivated to foster a better relationship with our parents, but it will take self-direction to plan and execute objectives like regular Sunday afternoon visits.

If, as we claim, boredom is rooted in ineffectual, objectless desire, then any failure of self-direction is, by definition, a key cause of boredom. People who chronically struggle with self-direction will be more likely to find themselves in the bind of ineffectual, objectless desire. Indeed, several dimensions of self-direction have been linked to the tendency to feel bored.[43] In our lab, we have examined the different strategies people use to achieve their objectives. Some people are focused on action and change—they prefer to "just do it." Others are focused on methodically and exhaustively evaluating the best course of action—they prefer to "do the right thing." It is clearly the latter who are more prone to experiencing boredom. They may get stuck ruminating about their current circumstances and fail to launch into some engaging activity.[44]

The causes of boredom within us, rather than facets of the world around us, describe well the boredom-prone person. Indeed, some of us seem fated to suffer more at the hands of boredom because of our psychological make up. Emotional, biological, cognitive, motivational and self-control factors all play a role. Combine these internal causes with the external causes we spoke of earlier and you have a truly toxic mix.

All Roads Lead to Boredom

It is not accurate to say that boredom originates in the environment *or* in us. It's both, or perhaps more precisely, neither. Boredom emerges from the way we come together with the world. The title of this chapter points to an unreasonable utopic expectation that boredom demands, that we could somehow cultivate a Goldilocks world where what we bring to the situation always fits perfectly with what the situation offers us. That is clearly not possible all the time. But when there is a strong mismatch, boredom lurks.

In saying this, one thing should be clear—there is no single cause, whether it is within us or within our environment, that can shoulder all the blame for boredom. Nor can we say that all causes need to be present at once for boredom to arise. But, during any given episode of boredom, if we follow the path back from our unoccupied mind and the desire conundrum, we will find one or more of these nefarious culprits at the source.

One way or another, the various causes of boredom do their work by preventing us from becoming engaged, and this strikes at the heart of what it means to be the authors of our lives—to be effective agents. Boredom's causes thwart our agency and, at its root, boredom is a crisis of agency. A cork floating in the ocean, pushed this way and that by the tide, is not an agent. The angler rowing his boat against the current to reach the shore is an agent. The cork is not determining its movement; the angler is.[45] The cork has no intentions; the angler wants to get to shore. Boredom tells us when we have become the cork. Controlling our mind, choosing what we will pay attention to, and then successfully devoting our mental capacity to that chosen focus is foundational to our sense of agency. That is the core issue in need of fixing when we're bored—we must reclaim our agency. We need to stop being the cork and become the angler. Boredom is that call to action.[46]

CHAPTER 3

THE MOTIVATION TO CHANGE

• • •

The children are bustling with excitement. Hundreds of people swarm along the shore of Lake Michigan on a perfect spring day—blue skies, light breeze, and the promise of better days ahead. The year is 1933 and Chicago is alive, celebrating a "Century of Progress!"

You stroll past multicolored tents and unimaginable exhibits. A Lilliputian city of midgets, exotic wild animals from all corners of the globe, babies in incubators, automobiles from the future! But your children drag you by the hand to the most bizarre tent of all—Ripley's Believe It or Not Odditorium!

Just taking the stage is an unassuming man—no taller than you, with neatly cropped hair and wearing white breeches, held tight by a black belt with a large brass buckle. The announcer, sporting a tuxedo, top hat, and cane, completes the circus atmosphere. He booms, "Behold, Ladies and Gentleman, a sight not for the faint of heart. Arthur Plumhoff—the Pain Proof Man!"[1] The audience is rapt. "An ordinary man with extraordinary powers. Watch as he repeatedly pierces the flesh without flinching once!"

Plumhoff steps forward with the gait of a street fighter. Without a word he picks up a five-inch needle. Crouching slightly, unblinking eyes staring, Plumhoff slowly opens his mouth, a gaping expression evincing more terror than awe. The room is silent. Theatrically he proceeds to push the needle first through one cheek, clearly visible through his open maw, before continuing on to penetrate the other.

Gasps of surprise and revulsion fill the tent. Your older son edges forward in his seat, enthralled. His younger brother cringed before the first cheek was breached and has cowered against your chest, eyes covered ever since. You think to yourself, "How is this even possible? How can a man stand such pain?"

• • •

"The Human Pincushion"—a man whose lack of pain led him to a life in the circus—was first discussed in medical journals in the 1930s.[2] His disorder—congenital analgesia—starkly highlights the function of pain: it alerts us to the need for action.

It's important to note that the purpose of pain is not to *cause* hurt. The *function* of pain is to signal the need to act. From the reflexive withdrawal from fire (Figure 3.1) or the retraction of your hand from an errant chop of the kitchen knife, to more deliberate actions like taking painkillers for a headache, pain signifies a need to do something to remedy the feeling.

This functional account of pain is not new. Pain has long been seen as an experience that interrupts our current focus of attention and motivates action to escape the painful experience; this disruption to our ongoing goals can, in turn, create an urge to restore the goals we had been pursuing before the pain began.[3] For physical pain, where tissue damage is the prime mover, this picture is relatively simple. But what of the kind of psychological distress we suggest is strongly associated with being bored?

Figure 3.1. René Descartes, *The Path of Burning Pain* (1664). This drawing is intended to show the neural pathways for a withdrawal reflex. The fire leads to a painful sensation, presumably burning skin, which is transmitted to the brain, where an action is planned. In this rather quaint drawing, the cherubic subject, sporting a bemused grin and showing little to no signs of pain, reaches calmly toward the afflicted toe. The fire is far more likely to elicit a spinal reflex than a deliberate action. The point, however, is that the pain generated by the fire demands action.

One theory suggests that pain, including psychological pain, operates as a self-regulatory signal.[4] While physical pain may elicit an automatic response, like removing your hand from the flame, psychologically unpleasant states like sadness—and boredom—may evoke more complex reactions. The distress we feel at the loss of a loved one evokes many responses, from wanting time alone to actively seeking the comfort of others. However we choose to respond to a psychologically stressful event, the point remains the same: pain—physical or psychological—operates as a signal, prompting action. We believe the boredom signal can be understood in much the same way.

Let's imagine an average day in the life of an office worker. Perhaps the day starts with some enthusiasm for what lies ahead. Our office worker is fully engaged in the tasks at hand. But it's hard to imagine being "switched on" all the time. The office worker must fight against distractions and deal with other signals to act, like a rumbling stomach, which communicates the need to eat. Sometimes, even the best attempts to stay tuned-in will fail. As time marches on, and things seem to be taking longer than expected, perhaps our office worker starts to fidget.[5] She pushes her chair back from the desk, stretches, takes a deep breath, and gets back to the grind. A little later on, she's checking emails, deleting spam and fidgeting a little more than before. A brief glance out her window turns into a five-minute daydream. Each of these episodes might reveal boredom is lurking, as she tries to lose herself in distractions, but perhaps it's not until the signal is strong enough that our office worker realizes that she's bored! Off-task and unsatisfied, she checks social media to see what her friend Bob is having for lunch today. That leads her to an article on climate change deniers from Sheryl's post and, just like that, she's down the rabbit hole of infotainment. The climate change piece leads to yet another article highlighting the latest gaffe by a prominent politician, and that leads to a fluff piece on pandas donated to the Toronto Zoo, which finally leads her to the sports page for a depressing recap of the local team's latest losing streak. At each point, she is disengaged and unsatisfied. The boredom signal tells her as much: This activity is not satisfying; do something else! There are many reasons why our office worker may have become disengaged—the inexorable march of time (things often become less interesting the longer we do them) or the sense of being trapped (having to do the task but not wanting to do it) might lead to sounding the boredom alarm. Whatever the reason, it is the boredom signal that highlights her lack of engagement.

Understanding boredom as a signal to act forces an important distinction between *engagement* and *meaning*, a distinction we will return to often. Although it's not popular to admit, most of us can imagine doing something deeply meaningful, let's say playing with our children, that nonetheless could be suffocatingly boring. Children seem far more content to persist in the telling of knock-knock jokes ad nauseam! On the other hand, you can imagine being fully engaged with something that you wouldn't normally say was particularly replete with meaning. Binge-watching the latest inane reality TV series may seem a waste of time in retrospect, but it was totally absorbing at the time. In this sense, the boredom signal tells us less about the content of what we are doing and more about the fact that whatever it is, we are not fully engaged by it. Turn the knock-knock session into a wrestling bout and maybe boredom can be avoided.

Another way to think of the boredom signal is to ask what a life without boredom might look like. We might even long for a boredom-free life. But on closer examination we should be careful what we wish for. A life without boredom would be filled with apathy and stagnation. At first glance, the claim that a boredom-free life would lead to apathy might seem crazy, because boredom and apathy seem so similar. Indeed, they both share the state of being disengaged. But as psychological states, they are, and feel, fundamentally different. The apathetic person is free of the pressured desire to find engagement. By its very definition, apathy is an absence of any desire to even bother redressing a lack of challenge—a failure of motivation, classically embodied in the couch potato. But for the bored person, things are very different. They are acutely aware of their strong desire for engagement—all of which leads to discomfort when that desire goes unmet.[6] As with pain, boredom motivates us to act—to redress the negative impact of disengagement. Eliminate that motivation and we may have a life without boredom, but in its extreme this becomes a life without desires of any kind.

Similarly, a life without pain might seem like something we would all want. However, for the "Human Pincushion" this meant a life fraught with the dangers of inadvertent self-harm. Perhaps attractive in the abstract, a life without boredom could lead to complete stagnation and a level of inaction that would ultimately be harmful to us as individuals and as a society. Our existence would have been short-lived if we did not engage our skills and talents to achieve goals. Imagine where we'd be if our ancestors had been content to remain mentally unengaged, lolling around the campfire (assuming they had the initiative to tame fire in the first place), never feeling the motivation to explore, create, and understand. No doubt, this would be a recipe for an unproductive and short life. Like pain, boredom is an important signal telling us we need to act to fully realize our potential.

The signal functions well when we respond to it *adaptively*. You might accurately read the boredom signal, but how you act next is critical. You could choose to engage in behaviors that are ultimately harmful—couch surfing behind your buddy's 4×4 or indulging in more than a few pints at the local pub. Or you could choose more productive outlets for engagement: go for a run, grab a new book to read, pick up your guitar and belt out a few classics. Unfortunately, time and time again boredom has been linked to maladaptive responses, including increased impulsivity and addiction.[7] Adaptive responses to the boredom signal need to invoke self-regulatory mechanisms to choose something more engaging than whatever we have been doing and to avoid potential distractions (in other words, leave social media alone until our lunch break). Succumbing to distraction and waning self-control represent potential antecedents to extended boredom. Chronically adopting maladaptive boredom remedies—from the extreme of couch surfing to the more mundane time sink of Facebook surfing—are unlikely to prevent boredom episodes from punctuating our days in the future, even though they may alleviate the state temporarily.

Consider boredom, then, as both a transient state and as a disposition.[8] We all know people who claim to never struggle with boredom. They may even state that "Only boring people get bored!" If indeed some people are blessed with a boredom-free life it might be because they are simply better at quickly responding to the signal before boredom becomes protracted. In this light, boredom only becomes problematic when it happens often and when our responses to it are ineffective or maladaptive.[9] As a transient state, boredom is aversive and disruptive, but in many instances may be relatively easy to remedy. Perhaps, on reading the signs of impending boredom, the boredom signal gets the office worker started on long-term project number two, immediately revitalizing the work day. Or perhaps she *does* become engaged by Bob's culinary posts on social media. It may be that a failure to respond appropriately to the boredom signal is associated with a higher frequency of experiencing the state and a difficulty in extricating oneself from it.[10] Those failing to respond to the boredom signal effectively may struggle to select an appropriately engaging new goal. Any failure to swiftly select something to do—adaptive or otherwise—may doom them to the prolonged, aggressively dissatisfying experience of boredom. Again, the distinction between content and process is important here. The boredom signal itself does not do any of the work needed to figure out what to do next or why what we're doing now is not enough. Determining what might be more engaging, meaningful, or satisfying likely depends on complex psychological processes related to motivation, reward, learning, and past experience. Boredom can't solve itself. It merely raises the alarm (Figure 3.2).

Negative feelings normally signal the presence of something relevant in our environment. A snake on the path ahead or an angry person rushing toward us, for example, signify an important event demanding a response. We suggest that boredom operates in much the same way. Where fear and pain signify the presence of something demanding a response—the snake on the path ahead stops us in our tracks, forcing a rapid search for a safe escape route—boredom signifies the absence of something; an absence of engagement. But how that absence is felt is a complex matter. Perhaps for some the onset of boredom is accompanied by physical sensations, such as a need to fidget, to pace, to dispense with unspent energy. For others there may be no discernible physical signature, but there may be a psychological need, poorly defined but ultimately arising from a feeling

Figure 3.2. Calvin, the perpetually daydreaming hero of the *Calvin and Hobbes* cartoon strip, detects the boredom signal successfully . . . but he is not thanked for externalizing the boredom alarm! Calvin's predicament highlights another factor important to the experience of boredom—situational constraints. He is trapped in the boring classroom with no avenue for an adaptive response.

that potential is somehow being wasted. Regardless, this absence of engagement poses an immediate challenge. I'm bored. Now what?

In this light, boredom need not be seen as entirely negative.[11] Although a negative feeling, boredom operates as a positive signal in at least two ways. It tells us that what we are doing now is no longer engaging, and it reminds us of our goals. It may even remind us that there are different, potentially better goals to strive for. Andreas Elpidorou, a philosopher from the University of Louisville, describes this as the "push" that boredom provides. Like pain, which motivates us to protect ourselves, boredom motivates us to seek out and engage with something that is perceived as more challenging, more engaging than whatever we are doing now.

Without the potential for boredom, the motivation to engage our cognitive capacity, we would squander our resources and fail to realize our potential. Contrasting tales of an eleven-year-old valve operator from the early eighteenth century and a pathologically lethargic teenager from the mid-twentieth century show just how important the desire to be mentally engaged really is.

Poor old Humphrey Potter's job was to repeatedly open and close valves at precisely the right moment to successfully operate Newcomen's atmospheric machine.[12] Monotonous doesn't even begin to capture it! Humphrey hated being a plug man. Even by 1713 standards, his was an exceedingly boring task. Watch for the right moment . . . open valve A . . . wait for the next critical moment . . . close valve B. Lather, rinse, repeat. Humphrey was bored and restless. He knew there had to be a better way. Indeed, there was, and unlike his more compliant fellow plug men, Humphrey was driven by primordial boredom to discover it. Humphrey noticed that he, as a thinking, decision-making, sentient person, had been rendered superfluous by his job. Valves were to be opened only when another mechanism was in a specific location and never at any other moment. So Humphrey set about devising a system of cords and gears to make the atmospheric machine do the work for him. Eureka! His boring job eliminated, Humphrey could now run off and play with the other children. Faced with extreme boredom, he had contributed to a monumental evolution of the steam engine—the introduction of the skulking gear—"skulking" being the eighteenth-century word for shirking work!

Flash forward more than 200 years to meet the yin to Humphrey's yang, Elsie Nicks. Elsie had suffered from terrible headaches for as long as she could remember. They got so bad she was prescribed morphine just to cope. But things were about to get even worse. As a teenager in 1941 she began to have "episodes."[13] She would become drowsy and extremely apathetic. She spoke only occasionally and in whispered monosyllables. Her behavior was far more reclusive than the cool, detached attitude common to teenagers. Eventually Elsie lost the capacity to act at all. This was not catatonia or paralysis—she simply had no drive. Elsie's case was a medical mystery of sorts. Was this an instance of the epidemic that killed almost a million people twenty years earlier—encephalitis lethargica?[14] Eventually, her Australian-born doctor, Hugh Cairns, discovered the culprit—a cyst in Elsie's brain rendering her inert. Draining the cyst resulted in brief returns to normal behavior. Ultimately, Cairns was forced to remove the offending cyst, restoring Elsie's capacity to act based on her wishes, wants, and desires. Cairns named her condition akinetic mutism.

There is something essential about boredom embedded in Humphrey's and Elsie's stories. Namely, boredom is an expression of the fact that we have intentions, and intentions are critical to functioning successfully in the world. Humphrey was committed to acting in a way that was engaging and used his abilities to their fullest. Because of this he was burdened by boredom when faced with the job of plug man. His boredom drove him to find a better way. Elsie, on the other hand, lacked the ability to commit to any course of action.[15] Because of this she was not burdened by boredom when faced with sitting for hours with nothing to do. Indeed, she could not even will herself to be dissatisfied because she did not want to do anything else. In a nutshell, someone with akinetic mutism shows a complete lack of self-initiated action. The condition is defined as an inability to form and maintain desires.[16] Essentially, Elsie became like a machine—able to behave only in ways others programmed her to. In the absence of others' commands, Elsie was content to sit for extended periods of time doing nothing. Apathy of this kind is in many ways the opposite of boredom. When apathetic we do not care, but when bored we care deeply. In fact, we are tormented precisely because of our desire for something satisfying to do, and we are bored precisely because that urgent desire goes unsatisfied.

Without the capacity to form intentions to act, without the drive to engage, we would also lack the capacity to experience boredom. We could consider that never experiencing boredom due to a failure to formulate any desires is a Pyrrhic victory.[17] Pyrrhus is the Greek general who lamented having won battles against the Romans at such an enormous cost that ultimately the war could not be won. Perhaps an absence of desires frees us from boredom in the moment (winning one battle) but ultimately prevents us from effective engagement with the world (in this analogy, losing the war, as Pyrrhus did). At their most foundational level, desires can be thought of as biological drives—drives that function to preserve our lives and the future of our clan. People that can't form desires are at serious risk of death. At their most lofty heights, desires represent human strivings that keep us searching for a better way, like Humphrey's need to escape the monotony of the work of a plug man. In comparison, Elsie couldn't even reach for a dropped candy, much less create a revolutionary gear to avoid the drudgery of work.

It's a double-edged sword. Having the capacity for desires, intentions, and projects puts us at risk for boredom, yet without them we would never innovate and develop as individuals or as a society. Humphrey's solution to his boredom was to create a machine—a machine that both relieved him from the drudgery of being a plug man and also resulted in a better steam engine. Machines, like Humphrey's skulking gear, don't shirk monotonous work. Machines, computers, and automatons will do the same repetitive task over and over again without complaint. In a manner of speaking we are advantaged by the possibility of boredom, whereas—from our point of view—the virtue of a machine is that it has no capacity for boredom. Indeed, machines that cannot be bored are a clever invention of biological organisms that can. Such machines have played an important role in our society. But, and here is where things take an ironic turn, we want more from our machines. We want our machines to be intelligent, and intelligence is a whole new ball game.

Think of your desktop computer. It may have incredible computational power, but no one would claim it comes even close to having what we would call true intelligence or the capacity for adaptive behavior. It can't even do simple things like make a cup of coffee. If we want to create artificial intelligence (AI) machines that will innovate and solve problems they were not originally designed for, we have to make them be more like Humphrey—prone to boredom. To make machines intelligent, we have to make them motivated to shirk monotonous work and driven to avoid squandering their computational resources. Evolution, it appears, clued into this long before AI researchers. But AI researchers are starting to catch up. And, examining the work that AI researchers are currently doing to make machines prone to boredom in order to be intelligent, we find strong evidence of our claim that boredom is a functional, adaptive signal.

An important aside is in order. Some philosophers and researchers suggest that mental states, like boredom, ought to be defined by what they do, not based on what they feel like or what is happening in the brain—this is the "boredom is what boredom does" school of thought. Functionalists of this stripe might claim that AI literally experiences boredom. This is not our approach.[18] Rather, we are sticking by what we said in Chapter 1, defining boredom by what it feels like and the underlying psychological mechanisms that give rise to that feeling. Here, our goal is to describe the functional role of boredom, not provide a new way of defining boredom. So when we say that intelligent machines must have the capacity to be bored in order to be intelligent, we mean they must experience states that play the same functional role that the feeling of boredom does in our own lives. We make no commitment to what machines are actually feeling, or even if they are capable of feeling states that approximate our own. However, this is not merely a case of anthropomorphizing. Rather, our excursion into the world of bored AI deepens our understanding of the function of boredom and bolsters our proposal that human boredom is adaptive.

A fascinating machine named Kismet is a good example. Like Humphrey, Kismet has what we might call desires, which make it both smart and vulnerable to boredom. Cynthia Breazeal, a professor at the Massachusetts Institute of Technology (MIT), created Kismet, an AI system intended to be capable of complex, humanlike social interactions.[19] Informed by research on facial expressions and the social difficulties seen in autism, Kismet is essentially a robotic head capable of facial expression that interacts with people.

Breazeal understood the importance of intentions for intelligent, self-directed behavior, so she gave Kismet three basic drives. Kismet wants to socialize, play, and rest. Kismet's drive to socialize motivates it to seek out and interact with its human companions. Kismet's drive to play motivates it to seek out and interact with toys. Kismet's drive to rest motivates it to quiet down and, if alone, sleep. It could be said that each drive specifies a set of needs that Kismet is motivated to satisfy. Like Goldilocks, however, Kismet will only flourish when it has the "just right" amount of each of these things—not too much and not too little. This idea of maintaining the "just right" amount is what biologists call homeostasis. All living things try to keep their internal conditions in an optimal zone. Our bodies, for example, work very hard to keep our core temperature relatively constant—the Goldilocks zone for our well-being. Disaster ensues if our core body temperature increases or decreases too much. Similarly, Kismet possesses homeostatic regulatory processes to maintain just the right amount of socializing, playing, and resting.

Looking more closely at Kismet's drive to socialize illustrates how this works. We've all got that one relative we try to avoid at Christmas dinners, right? An intrusive aunt who stands too close and never stops talking about herself? You instinctively want to move away, excuse yourself from the dinner table to—supposedly—use the bathroom. When faced with this kind of intrusive aunty, Kismet does something similar. It averts its gaze in an attempt to break off the interaction. In essence, aunty has overwhelmed the AI, and its socialization bucket is overflowing. On the other hand, we know what it's like to be cooped up at home in the depths of winter for days on end with no one to talk to. Cabin fever strikes. You might even get desperate enough to give your aunt a call! This is what too little social contact feels like. Kismet's motivation to socialize increases to urgent levels in these instances. Its socialization bucket is empty, and it is desperately trying to catch the attention of a human in the room, expressing the need to communicate. So whether from too much or too little interaction, Kismet's motivation becomes more intense as its level of socialization deviates from the "just right" zone. These changes in motivational intensity are associated with different emotional states. "Just right" is comfortable. Too much is distressing. Too little is boring. Kismet's capacity to *want* gives rise to the possibility of it being bored. The yearning for satisfying engagement sets the conditions that activate boredom within Kismet. This state of boredom pushes Kismet to explore its surroundings in order to find things that will satisfy its basic drives.

Babies come into the world motivated to explore and influence the world around them. This motivation turns out to be the critical foundation for development. Without it, learning would not happen. In fact, it could be said that we become intelligent by exploring things around us. In line with the notion that exploration represents a foundation of human development, Alan Turing proposed that the way to create true AI would be to simulate the mind of a child.[20] Pierre-Yves Oudeyer noted that the key in Turing's advice is that AI researchers should solve the problem of *why* a robot would learn, rather than only thinking about *how* they learn. That is, we need to give robots the desire to explore and manipulate their environment and this, in turn, will give rise to intelligence. This is exactly what Oudeyer and his colleagues have been doing—designing robots that are intrinsically motivated to investigate their surroundings. He then puts these robots in situations rich with opportunities to explore and manipulate, stands back, and watches how they become competent through interacting with the environment. He also found that when in highly familiar situations that do not offer opportunities for *novel* engagement and learning, the robots get bored. Once again, we see that boredom is a price that robots—like us—have to pay if they want to be intelligent. In this light, boredom and motivation are inextricably linked.

Jacques Pitrat came to the realization that AI systems must have the capacity for boredom through working with his co-investigator CAIA.[21] CAIA—otherwise known as Chercheur Artificiel en Intelligence Artificielle (or Artificial Researcher on Artificial Intelligence)—is both Pitrat's creation and able assistant. In a clever move, Pitrat decided to study AI by creating an AI researcher. Now, almost thirty years in, CAIA is valuably contributing to his efforts. As a bonus, Pitrat can observe CAIA in action and learn from its performance. Pitrat's goal is to eventually create a completely self-directed, artificial AI scientist. To achieve that goal Pitrat gave CAIA the capacities of self-observation and self-evaluation. Armed with the capacity for self-observation, CAIA can notice what it is doing and understand why it failed or succeeded in solving a problem. With the capacity for self-evaluation, CAIA can prioritize the easiest and most important problems, determine if a problem is even worth attempting to solve, and decide if its solutions are useful. Combining these capabilities, CAIA is able to pull out of unproductive computational loops. Essentially, like Humphrey Potter, CAIA is motivated to avoid squandering its resources—and thus, you could say that CAIA tries to avoid boredom.[22]

Unfortunately, like the hapless child pleading with their parent to alleviate their boredom for them, CAIA can't do anything about its boredom other than stop and complain. CAIA signals its boredom and then waits for Pitrat to step in and solve the problem. In turn, learning when and why CAIA becomes bored provides Pitrat with critical insights toward his goal of creating better AI systems. Ultimately, CAIA may one day be able to solve its own boredom through innovation and, like Humphrey, invent a skulking gear.

Kismet, on the other hand, has a rudimentary mechanism to alleviate boredom. The key to Kismet's ability to alleviate boredom is something called a "getting sick and tired of the same old thing" (habituation) component to its programming. This component causes a shift in Kismet's "attention" when it is stuck with the same old, same old. Like Humphrey and CAIA, Kismet knows when to quit. Rodney Brooks, the head of Breazeal's lab, calls this the "Steven Spielberg memorial" component in honor of the robot in Spielberg's movie *AI* who sat, unproductively, staring at a statue for 2,000 years! Brooks and Breazeal wanted to ensure Kismet would not make the same mistake, so they gave it the capacity to be bored.

Boredom can strike when things don't ever change, and Kismet is programmed to react to that. But boredom can also strike at the other end of the extreme, where things are constantly changing. An ever-changing environment is too chaotic and noisy for us to make any sense of, ultimately leading us to disengage out of boredom.[23] This too has been a problem for AI researchers. In work intended to imbue AI agents with the drive to be curious, researchers found that their agent—which was learning to navigate a virtual maze—got stuck in the maze if they placed a virtual TV screen with continuously changing content on one wall.[24] Each new image on the screen was novel, satisfying the agent's curiosity drive. Clearly, getting stuck in one spot indefinitely would not be an adaptive thing for humans or AI, certainly not when the goal was to explore and learn to solve a maze. Interestingly, in other computational work, boredom was shown to be a better driver of exploratory behavior than curiosity. When researchers created two distinct types of artificial learning agents—one driven by boredom and the other by curiosity—it was the bored agent that learned best.[25] This is not to say that curiosity fails to drive exploration and learning. But the bored agent would not be captured by the TV with constantly changing content. It would eventually get bored by the meaningless noise and move on, while its curious cousin would stay glued to the idiot box.

CAIA and Kismet can get bored and quit. In his provocative book *The Dip: A Little Book That Teaches You When to Quit (and When to Stick),* Seth Godin highlights the value of what he calls "strategic quitting" to avoid getting stuck in a cul-de-sac of wasted effort.[26] Counter to the famous saying, "Winners never quit and quitters never win," Godin notes that winners quit all the time, and knowing when to quit—before you have squandered your resources—is a really valuable skill. Boredom—the motivation to change—is our ally in this regard, pushing us to move on to something new. Elsie lacked it, Humphrey had it, and smart machines are getting it. We should not be so quick to ask for a boredom-free life—just ask the Human Pincushion what happens when you can't experience pain.

Pain of any sort—emotional or physical—feels bad; we don't like it and want to be rid of it as soon as possible. It's the same with boredom. So the most immediate and pressing message of boredom is to get rid of this horrible feeling. And indeed it's good to make a change. Being mentally unengaged is of no good to anyone. The sixty-four-thousand-dollar question, however, is: What *should* we do? Boredom can't directly answer that question. Maybe we should redouble our efforts with the task at hand so as to become engaged. Maybe we should try our hand at something else. But if so, what? To compound our predicament, boredom feels so bad that we are driven to grab the quickest, easiest, most soothing balm, which may not be the best for us in the long run. So to hear the deeper message behind "get rid of this horrible feeling," to respond adaptively to the motivation to act, we must keep in mind the causes of boredom we discussed in the previous chapter. It's the same with physical pain; we can only respond in a truly effective manner when we understand the cause. Otherwise, we are left haplessly trying out one possibility after another and believing we have found a solution that might only be a temporary Band-Aid.

At its deepest levels, boredom tells us we are squandering our abilities and not engaging with the world in a way that fulfills our agency or, as Robert White put it, our need to express and develop our competence.[27] So, when feeling bored, the message is not merely to make the disagreeable feeling go away as fast as possible, but to find a way to interact with the world that better engages you and gives expression to your desires and abilities. For boredom's call to action to be effective, we must keep in mind that below the superficial motivation to change, boredom is ultimately the motivation to express ourselves as agents, in control of the choices we make. At its best, boredom won't let us rest until we take up our agency, discover what we desire, and cultivate engaged interactions with the world.

But this is tricky. In the moment of being bored we can feel as though we lack control, to the point of feeling all we can do is complain. Yet, it is precisely in these moments that we most need to rediscover our agency rather than to treat ourselves as a vessel to be filled, titillated, or soothed. Being an agent is something we have to work at. We have to take steps daily to foster agency, both in ourselves and in those around us. And we have to be on our toes, because the forces that thwart our agency can change over time. At each stage of life, the causes of boredom manifest in different ways.

CHAPTER 4

ACROSS THE LIFE SPAN

• • •

The store security officer would have been surprised had it not been the third time this month. Here she was again, sitting in the store manager's office, crumpled and forlorn. The things she'd been shoplifting didn't even make sense. One time it was a breast pump and baby clothes, and she was clearly not pregnant and highly unlikely to be so anytime soon. This time it was a pair of boots at least two sizes too big. He had come to expect random thefts from teenagers—that was the norm. Kids had too much time on their hands and nothing to do with it. But June was 76 years old and a great-grandmother!

"Answer just one question for me, June?" So many incidents had put them on a first-name basis. "Why do you do it?"

The pensioner looked up at him with an expression halfway between disgust and boredom.

"Do you know what it's like getting old? I'm bored. There's nothing to do. That's all."

There you have it, he thought, not so different from teenagers stealing candy bars.[1] Her desire to steal arose from a need to do something.

• • •

Teenagers and septuagenarians are both deeply uneasy when they have too much time on their hands and nothing to do with it. We're not built to have endless hours of wandering through a mall when school is over. And we're not built to sit at home and watch game shows and soap operas all day to fill out our retirement years. Boredom at the bookends of life, from cradle to grave, signals to us that we need something more.

To date, boredom has been examined through a fairly narrow lens in terms of age. By far the most commonly tested age group is 17- to 22-year-olds—undergraduate students at universities around the world where most of this research happens.[2] In this group, boredom propensity declines with age. But does this mean that there is a steady march toward lower boredom as we age or are there peaks and valleys? Are there predictors early in life that could tell us who will be more or less prone to boredom? And does boredom at the end of life have the same causes that it does at the beginning?

One of the first studies to say anything about boredom across the life span was actually more interested in one of boredom's opposites—curiosity. Leonard Giambra and colleagues from the National Institute on Aging in Baltimore examined curiosity and sensation-seeking across the life span. As a corollary to these experiences, his group also looked at the tendency to feel bored.[3] They found, as we and many others have, that boredom levels decline in the late teen and early adult years. But they went beyond the early twenties and saw continued decline into the fifties (Figure 4.1). Beyond the sixties, however, boredom levels started to gradually rise again, particularly in women.

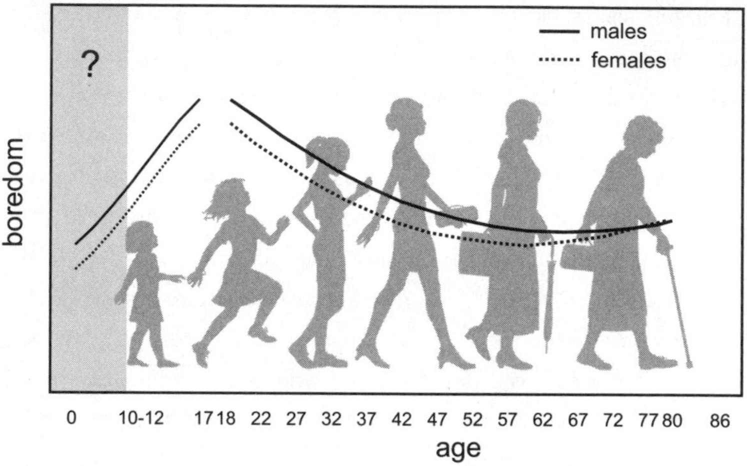

Figure 4.1. This graph was created using data from a 1992 study by Giambra and collegues and a number of short time-scale studies that span the range from 10–12 to 17 years of age. The Giambra data show a quadratic relationship between age and boredom such that boredom decreases from the late teens before rising subtly after the early sixties (60–65). The rise in boredom between the ages of 10 and 17, and the decrease from the late teen years is well studied—far less is known about the later decades of life. We know next to nothing about boredom before age 10 (the gray area with a question mark).

This changing relationship between a tendency to boredom ("boredom proneness") and age invites a fascinating question: Is boredom changing across the life span for specific reasons? We note that the decline in being prone to boredom in the late teens and early twenties parallels the final stages of neural development. This might be one explanation for the observed changes in boredom. That is, as the frontal cortex comes fully online, the propensity to experience boredom declines. The subtle rise in boredom at the other end of the age spectrum may also be related to the frontal cortex—this time due to a decline in function within this brain region that accompanies normal aging.

Circumstances too likely explain the relationship between age and boredom proneness. At precisely the time that the frontal cortex is reaching full maturation, most countries also imbue their citizens with a swathe of rights, responsibilities, and freedoms.[4] Being able to drive, vote, join the army, consume alcohol, all signal a change in our environment and our own capacity to interact with that environment that gives a greater sense of agency and leaves less room for boredom. There is simply more opportunity for self-determination, more stuff we can do to stave off boredom. And what of middle age? Here, a different set of responsibilities descends on us that could simply preclude the experience of boredom for most. With careers, spouses, children, and mortgages there may simply be less opportunity to feel bored in our middle decades. Retirement brings with it a release from many of those responsibilities. But if the environment or our own physical and mental limits do not allow full utilization of our abilities, we risk becoming disengaged and isolated (Chapter 6), and boredom may rise again.

When it comes to explaining changing levels of boredom proneness across age we are largely left to speculate, as there is very little research on the topic, which is somewhat surprising given that informal observations about age and boredom have been made for quite some time. We simply do not know for sure what biological or social factors differentially impact boredom levels across the age spectrum. There is some research looking at the impact of social factors *within* younger teenagers with a particular focus on specific circumstances (such as education settings) or demographics (such as rural versus urban settings). But that is about it. And none of this touches on broader cross-cultural differences. Do the stark differences between individualism in the West and collectivism in the East mean that boredom manifests differently across the age spectrum? Even more subtle differences of culture—between the United States and Canada, for example—might influence how we perceive and respond to boredom. Clearly, we have a lot of work to do.

"I'm Bored!"

Anecdotally all parents know well the experience of their children pleading with them to remedy their ennui. Most of us simply dismiss boredom in children as *their* failing. Given our description of boredom, dismissing it as trivial and suggesting multiple avenues to remedy the boredom clearly misses the point—boredom does not arise from a lack of things to do, or as a consequence of diminished motivation to engage. Just the opposite. Bored children know there are many things in the world to do and clearly want something that will satisfy them. What they are struggling with is the "how to" of that equation. The fact that they demand we fix it for them may simply reflect the reality of their surroundings. As their parents we control much of their world, so why not this part too? Unfortunately we may be predisposed to tune them out when they're bored. It turns out that oxytocin—the bonding hormone—triggers an increase in empathy to crying children only when we think their cry indicates that they are ill. When we deem their wailing to be driven by boredom, our oxytocin remains silent and we are unmoved.[5] So bored children are left to figure things out on their own—which, in the long run, might actually be best for everyone involved.[6]

Childhood boredom, routinely dismissed by parents, has also been overlooked in the research world. Understanding boredom in childhood, and indeed across the full spectrum of age ranges, is hampered by design constraints that all studies interested in changes over time face. Ideally, we would conduct longitudinal work—start looking at boredom in the preschool years and follow the same individuals for as long as we can. It's unlikely that we could follow them all the way to the end of their lives, but even trying to capture a five- to ten-year range is challenging. Given this challenge, we are left with what are referred to as cross-sectional studies—exploring boredom in different groups of individuals at each age range of interest.

Studying boredom in very young children presents other challenges. When 4-year-olds say they are bored, are they using the word in the same way we are? Beyond that definitional issue, even using the word *boredom* may be seen as taboo in classroom settings where young children spend most of their time. Whatever the challenges are, the consequence is that little has been done to understand boredom in anyone younger than 10 years of age. In one study researchers asked third- and fourth-graders about their experience of boredom and related their responses to math and reading abilities.[7] Boredom was associated with poor academic achievement, but it was most strongly related to reading. Kids with better reading skills reported lower levels of boredom. Reading demands imagination to turn words into images, dialogue into imagined accents, scenes into dynamic mental events. Better reading skills then, may reflect higher levels of imagination, which in turn reflect a stronger engagement with the material. We would argue that effective engagement with whatever you are doing is critical to stave off boredom. Grades three and four mark an important transition from learning-to-read to reading-to-learn. If a child is struggling to read, it becomes more and more difficult for them to engage with classroom activities.[8] Unfortunately, studies like this one represent a snapshot in time. They don't tell us what factors might explain changing boredom levels with age.

Boredom has long been associated with increased sensation seeking and risk taking in adults. In children, Mary Russo and colleagues from the University of Georgia suggest there are modest increases in sensation seeking beginning around age 7 that continue into early adolescence.[9] This, combined with limits in self-determination, is potentially a recipe for rising boredom levels. On the one hand, a child wants to seek and experience new sensations. On the other hand, their capacity for self-determination—not just to choose what they want to do, but to execute their desired goals accurately—is limited first by their own developing cognitive and physical capacities and second by the external constraints imposed on them by parents and society.[10] This clash between abilities and the constraints of the world becomes even more evident as we move from childhood into adolescence.

The Rising Tide

On North Baffin Island, 11- and 12-year-old Inuit boys accompany their fathers into the wild to hone hunting skills. Girls in many cultures take part in rituals tied to observable biological changes, often coinciding with their first experience of menstruation. Boys as young as 7 or 8 on the island of Vanuatu jump from a tower nearing thirty meters high with their feet bound by a less than bungee-like vine to celebrate ascendency to adulthood. What each of these traditions has in common is a culturally defined transition from childhood into teenage years, often at a time when many of our more sophisticated cognitive abilities start to come on line.[11] This transition may also represent a key point in the expression of boredom across the life span.

It seems like a cruel joke: the more capable we become of exerting influence on the world, the more we open ourselves up to boredom. This stems from our assertion that boredom arises from failed attempts to engage with the world. As our capacity to engage expands with the development of new cognitive skills, so too will the *possibility* of finding our newfound capacities underutilized. And this will be especially true if the world does not yet afford us a full suite of options for action, which is true for most teenagers. Teens are faced with either restrictions on their possible behaviors or too much time on their hands with too little to do. Either way, boredom ensues.

Many of the rituals just mentioned follow the biological transition into adolescence far more closely than any particular chronological age. Along with the rush of hormones and heightened emotional intensity comes newfound cognitive abilities.[12] The newly minted teenager is on the doorstep, not just of increased emotional intensity and complexity, but of improved capacity for thinking in the abstract, for complex problem-solving, and for reasoning through challenges in a logical manner. It's what makes teenagers what they are—emotionally intense, cognitively flourishing, and infuriating to argue with.

Yet, developments in emotion and cognition within the teenager do not follow the same trajectory. On the one hand, the emotional networks of the brain experience rapid changes due to the flood of hormones, while on the other hand, the cognitive, reasoning system starts its gradual creep toward full adult capacity over the decade of the teen years and beyond.[13] As Ronald Dahl, a researcher at the University of Pittsburgh, suggests, this is akin to "starting the engines with an unskilled driver" behind the wheel.[14] Others have even suggested that these two systems, with their distinct developmental and functional profiles, are in conflict with one another—a kind of push-pull of emotion and reason.[15]

This model implies that increased sensitivity to reward and intense emotions, coupled with a developing but incomplete capacity to deal with these emotions, may be at the heart of rising boredom levels over the teenage years. All of this may also intensify sensation seeking. Developing cognitive skills require an outlet, and this drive may push the teen toward both curiosity and information seeking, as well as seeking out the thrill of experiencing new things.

This is a complex maze of circumstances the teen must navigate. They experience a strong drive to express themselves and explore the world, are unable to rationally corral and control their intense emotionality, and continually run smack into intransigent adults with their rules and limits. Teenagers must attend schools with class schedules decided for them. Parents determine most extracurricular activities and restrict time with friends, time spent on screens, and so on. These constraints operate to counteract a strong urge in the teenager—the desire for autonomy or self-determination.[16] Certainly, in one study of college students, the most prevalent term used to describe the experience of boredom was "restlessness," perhaps born of the desire to break out of the constraints imposed by the learning environment.[17]

The flip side to such constraint is having too much time with nothing to do, another hotbed for boredom. By some estimates, around 40 percent of adolescents' time is free time.[18] And for many teens this so-called leisure time is boring.[19] Teens with ample time on their hands, elevated levels of sensation seeking, and few outlets deemed likely to satisfy are underutilizing their skills and are prone to becoming bored.[20]

A recent study of youth in South Africa provides a good example; youth who had more free time also engaged in higher levels of sexual activity.[21] Boredom played a key role. In general, youth in this study who had lower levels of employment, lower socioeconomic status, and more free time reported higher levels of both boredom and restlessness. The study was one of the few that was longitudinal, tracking teens over two years of high school. Those who reported higher levels of boredom at grade nine were, the following year, more sexually active and more sexually aggressive. Males who reported higher levels of boredom in grade nine were also more likely to engage in *riskier* sexual behaviors, such as not using a condom or having casual sexual encounters. In a similar vein, a study of youth in rural New Mexico found that increased levels of boredom were associated with more unstructured time and limited opportunities for engaging in meaningful activities.[22] In turn, these teens also engaged in higher rates of drug use and were more likely to be involved in "trouble making." Clearly, being bored and having too much free time on your hands has consequences.

Risky sex, drug taking, and trouble making. All may represent a constellation of teenage desires: to assert independence, to seek new experiences, to fulfil the promise of newly developed skills. When pitted against an environment constrained through parental control, institutional restrictions, or a lack of opportunities, boredom thrives.[23] Then, just when the teenage years have hit the height of emotional turbulence, from age 17 onward boredom levels start to drop (Figure 4.1).[24] This drop comes at precisely the age when opportunities for autonomy and self-expression open up and at a time when the capacity for self-control starts to mature.

I'm an Adult Now

From an institutional point of view, 17- and 18-year-olds are leaving the school system either for employment or further education in environments—universities and colleges—rife with choice. They are bestowed with rights and responsibilities they never had, from voting and driving cars to legally consuming alcohol (at least in some parts of the world).

This explosion of freedom and opportunity at the tail end of the teenage years is likely to be only part of the reason boredom levels dip. As we've already suggested, the onset of puberty might signal a potential rise in boredom, independent of any specific chronological age. Here, in the later teen years, chronological age may once again be less relevant than the biological processes churning away in the background. By 18, as that complex emotional soup of human existence has been unfolding, teens have begun honing the cognitive skills needed to better act on desires and goals. There is still some way to go—the brain is not fully developed until the early to mid-twenties. But by the late teens, there is tangible progress in development of the part of the brain known as the frontal cortex.[25]

The frontal cortex of the brain acts as a kind of CEO for the rest—controlling complex behaviors based on information fed forward to the frontal cortex by brain regions dedicated to more basic sensory and motor processing.[26] Functions like abstract reasoning—being able to conceive of such things as beauty or bravery; planning ahead—the kind of future thinking that enables humans to chart out their career paths; and inhibitory control—being able to stifle a laugh at a funeral, all count as executive functions. Each of these things is considered an executive function because it is complex, multifaceted, and invoked in a seemingly voluntary, willful manner. In short, the development of the frontal cortex enables a greater degree of self-control and autonomy.[27]

Noting what happens when a person of any age experiences damage to this critical part of the brain bolsters our notion that the decreases in boredom seen in early adulthood are related to development of the frontal cortex. All too often car accidents, sports concussions, and barroom brawls lead to traumatic brain injury (TBI). Close to 2.8 million people are diagnosed with TBIs every year in the United States alone.[28] The parts of the brain most affected by this are the frontal cortex,[29] and the *sine qua non* of TBI is what experts refer to as dysexecutive syndrome.[30]

For some time now clinicians working with people who have experienced brain injury have noted that patients often complain of boredom.[31] Our own data show that in TBI patients self-reported levels of boredom are higher than in healthy individuals—something that until now we had only suspected to be true from anecdotal accounts and clinical experience.[32] For our patients this was not simply a response to the monotony of hospital life, given they had all long since been discharged. Instead, it seems as if something has been altered in people who have suffered a TBI that makes it more difficult for them to engage with their world in a satisfying, meaningful way.

It is likely that impaired levels of self-regulation and control represent the key changes that drive an increase in boredom among TBI survivors. Conversely, increased development in these functions in the late teenage years might be precisely what causes the rapidly diminishing levels of boredom just as we venture into adulthood. What is also clear from the data is that boredom continues to fall, even after full brain maturation is long behind us.

Middle Age, Middle Boredom?

Notably, during midlife, a time infamous for dissatisfaction and acting out—for men embodied in the ubiquitous purchase of an impractical sports car—people seem *less likely* to report boredom. Returning to the Giambra study (Figure 4.1), we see that boredom levels drop from the twenties onward, hitting a floor in the fifties before a slight rise in the sixties and beyond. This pattern of declining boredom during midlife was recently confirmed by Alycia Chin and colleagues,[33] who collected an enormous amount of experience sampling data from close to 4,000 Americans with an average age of 44. Consistent with Giambra, Chin found that older adults were less likely to report boredom. It's worth noting that the decline in boredom was not linear. That is, while the boredom levels of a 25-year-old were four times higher than those of a 45-year-old, the boredom levels of 45- and 60-year-olds were comparable to each other.

Thanks to Chin and colleagues, we also have some sense of what boredom looks like including in mid-life. First, they found that boredom did indeed occur at least once over a seven- to ten-day period for 63 percent of people enrolled in their study. Of seventeen different feelings reported, boredom made it into the top ten, at number seven. Of the *negative* feelings reported, boredom was the fourth most common, right behind exhaustion, frustration, and indifference. When it did occur, boredom was often associated with other negative feelings, such as loneliness (something we turn to in Chapter 8) and anger and sadness (something we will deal with in Chapter 5). So despite the fact that boredom rates continue to drop into our middle years, it is not as though boredom disappears altogether. It remains very much a part of our everyday lives.

Research by Chin and her colleagues may also provide some clues as to why boredom levels decline from young adulthood to midlife. The key might be how younger and older people spend their time. It turns out, based on this study, that people report boredom most often when they are studying, attending school or college, and associating with people they do not know. It is reasonable to think that younger people, rather than older adults, are more likely to find themselves in these circumstances. The notion that boredom is less common among older individuals because of how they spend their time was at least partially confirmed by Chin and colleagues' statistical analysis. In fact, they reached the more general conclusion that boredom is largely determined by the situations people find themselves in as opposed to differences between people, such as their age.

Truth be told, we know very little about midlife boredom, and this lack of research may itself indirectly implicate the role of life circumstances in boredom at midlife. That is, perhaps those in midlife are simply too preoccupied with building careers, starting families, and taking on responsibilities like mortgages to participate in research studies, and these same factors that keep them out of researchers' labs may also be responsible for the relatively lower levels of boredom they experience.[34]

What this contrast between young adults and those in the middle of life highlights is the influence of one's circumstances on the potential to suffer from boredom. Contextual influences stretch well beyond careers and mortgages. The impact of environmental context is evident in the elderly too, and this may converge with changes in cognitive skills to fuel a late-life surge in boredom.

Boredom in the Elderly

Throughout we have claimed that boredom arises when the desire to engage with the world goes unfulfilled, and we feel mentally unoccupied. For the elderly, two key culprits foment this breeding ground for boredom. First, as we age our cognitive capacity declines, giving rise to challenges of self-control and attention that we know are associated with increased boredom. Second, as we age our networks of social contacts decrease, as do environmental opportunities for engaging in satisfying activities. These two culprits likely conspire, each amplifying the effect of the other.

Ronán Conroy and colleagues from the Royal College of Surgeons in Ireland examined a group of people sixty-five years and older to figure out what was associated with cognitive decline in this population.[35] They found three key factors: *low social support* (people living alone with low external social supports), diminished *personal cognitive reserve* (low levels of social activity and leisure exercise, increased levels of loneliness and boredom), and lower levels of *sociodemographic cognitive reserve* (people living in rural communities and with lower levels of education).[36] Cutting through the technical jargon, it was loneliness and boredom that were associated with reduced cognitive function. Loneliness and boredom in the elderly may more generally represent a failure to optimally engage with the world. That failure may be driven, at least in part, by a decline in cognitive functioning. Certainly, research shows that cognitive decline in the elderly is predominantly evident in the area of planning and self-regulation, both executive functions—those same functions, normally supported by frontal cortex, that may keep boredom at bay in the late teens and early twenties.[37]

Tragically, just when the ravages of time degrade the cognitive function of many, so too time steals away opportunities for satisfying engagement with the world. Friends slip away, physical ailments limit our capacity to engage in ways we used to, and external circumstances become more restrictive—all putting us at more risk for boredom. The changes seen in later life can leave us under-challenged and under aroused.

Gillian Ice from Ohio University suggests that nursing home residents spend close to half their day doing next to nothing—sleeping, watching television, and engaging in other passive activities.[38] Clearly, these activities are underwhelming and limit residents' capacity to get energized about life. Indeed, the elderly report higher levels of boredom, which may be an obvious response to having little to nothing to do. Importantly, the elderly also report being restless and fidgety.[39] This combination of lowered physiological arousal and feelings of restlessness is common in the boredom literature throughout the life span, and not surprising to find among the elderly. It's hard to imagine raising one's heart rate by watching daytime television every day! Studies of teens consistently highlight the strange bedfellows of restlessness and lethargy in response to the perceived monotony of life, itself a direct consequence of feeling as though there is nothing to do. So, at the bookends of life, the mechanisms that cause boredom may be the same—a sense that we are underutilizing our skills and talents.

This feeling that we are not mentally occupied but instead engaging in pointless activities is closely tied to our sense of life meaning more generally. We'll explore meaning making and boredom later, in Chapter 7. For now, suffice it to say that institutional living may fail to provide enough meaning to ward off residents' boredom.

While pinning down how boredom manifests itself across the life span has been challenging due to the many gaps in research and the lack of longitudinal data, we can glean important themes. Biological milestones are critical: the transition from childhood to teenage years has emotional intensity colliding with developing but incomplete cognitive skills. Toward the tail end of our teenage years and into early adulthood, we see a transition to a more mature brain, one that is better able to exert control over our thoughts and emotions. Later in life a decline in those same capacities then becomes the culprit in the experience of boredom. Circumstances are just as crucial: teens with ample free time on their hands and few outlets for expression find themselves in much the same situation as the elderly living in institutional settings that afford little of consequence to engage with. The river that runs through it all? Whether we are 4, 40, or 80, boredom arises when we feel we are underutilizing our skills and talents. We know we could be doing more and we *want* to be doing more, but we can't seem to scratch that itch. Whenever it arises, the predicament that boredom signals is no trivial matter.

CHAPTER 5

A CONSEQUENTIAL EXPERIENCE

• • •

It ought to have been a routine flight. Northwest Airlines Flight #188 was scheduled to fly from San Diego to Minneapolis, departing at 5:01 p.m. central daylight time. Two hours into the flight, at 6:56 p.m., Air Traffic Control lost contact with the plane over Denver.[1]

Why were they not responding? Had something happened, accidental or otherwise? Thoughts turned quickly to worst-case scenarios, 9/11 never far from the minds of those in the industry. Contact was made with NORAD; fighter jets were prepared for take-off.

Tensions ran high everywhere but in the cockpit. For the pilots, the hard work was over: cruising altitude had been reached, and the plane was virtually flying itself. With nothing but an endless, monotonous horizon to look forward to, boredom set in. Their minds wandered. Restless, the captain left for a bathroom break. Not unusual, but while he was out Air Traffic Control asked the first officer to switch radio frequencies. He did so, but mixed up the numbers and got Winnipeg when he wanted Denver. The chatter on the radio from Winnipeg fooled him into thinking he had the right channel. The captain returned, and he and the first officer talked about a new scheduling system—anything to pass the time. Needing something to occupy their bored minds, they were both paying attention to their laptops. Winnipeg continued to chatter while Denver frantically tried to make contact.

Headquarters at Northwest joined the effort via an on-board data link system. In most planes this involved an audible chime, but not the Airbus A320—just an itty-bitty light, illuminated for thirty seconds, which was clearly not enough to capture the drifting attention of the pilots.

Finally, the error was detected; the pilots realized they were 100 miles off course. Air Traffic Control sent them on a series of turns to ensure the pilots were in full control and not operating under threat from a terrorist. The fighter jets were called off, and the plane landed safely, passengers presumably none the wiser.

• • •

Boredom gets blamed for all manner of ills. Acts of aggression, substance abuse, addiction, and an over-reliance on our smartphones have all been attributed to boredom. Yet, the true culprit might not be boredom itself but rather the company it keeps. For example, British civil servants who reported a great deal of boredom at the start of a survey were more likely to have passed away from heart disease three years later compared to those who reported no boredom—suggesting the possibility of literally being bored to death. Yet, when controlling for employment circumstances, health, and levels of physical activity in the study, the link between boredom and death disappeared.[2]

The study of boredom's consequences is fraught with challenges, and much of the existing research is simply not rigorous enough for definitive conclusions. One issue is the important distinction between the feeling of boredom in the moment (state boredom) and the personality-based tendency to feel bored frequently and intensely (trait boredom). It is one thing to say that the personality-based tendency to feel boredom predicts a psychosocial problem and quite another to say that the state of boredom causes a problem. We contend that much of the past research on boredom's consequences has not paid enough attention to this distinction. Throughout, we will use the term "boredom" when referring to the feeling state and phrases such as "the tendency to be bored" or "boredom proneness" when referring to the personality trait.

In this chapter we focus on possible consequences of boredom that are either supported by lab-based manipulations of state boredom or longitudinal studies that have shown that changes in boredom *precede* changes in the purported consequence. Some of these longitudinal studies have assessed state boredom, whereas others measured the level of boredom during an extended period of time. The British civil servant study, for example, asked about boredom levels over a four-week period at the start of the survey.[3]

Although boredom and the tendency to be bored may not *directly* cause everything they are blamed for, how we respond to the uncomfortable feeling of boredom is key. Boredom can drive us to act in ways that are maladaptive and directly impact our well-being. When bored we could choose to get started on that detective novel we bought months ago, or we could just kick back on the couch with a bag of chips and watch TV mindlessly for the afternoon. Clearly, one choice is healthier than the other. So it is in this indirect sense that boredom causes problems. Our responses to the boredom signal are the more direct cause of problems. But what are the consequences of failing to heed the boredom signal in adaptive ways?

As the story of the Airbus flight to Minneapolis highlights, when doing simple, repetitive, unsurprising, or familiar tasks, it can be hard to keep our eye on the ball.[4] In many situations this is not of any great consequence. Mowing the lawn is simple, repetitive, and all too familiar, but if we miss a spot on the lawn we can simply go over it next Sunday when the lawn needs mowing again. But if what we are doing has real consequences, like flying a large aircraft with hundreds of passengers, then our failure to stay focused can have dire outcomes.

The more bored you feel while doing a repetitive, monotonous, and simple task, the worse you perform.[5] Things that require your attention but that do not fully consume your mental capacity are boring. The way to get rid of boredom is to engage your mind. But, and here is the catch-22, the boring task by definition is not capable of satisfying your need to be mentally occupied. If you seek something else to engage your mind, you fail at the boring task—or at least, you won't perform at your best. Assembly-line work is now mostly automated for this very reason. Monitoring a line of widgets for quality control is monotonous and boring, making it hard to sustain attention and increasing the threat of succumbing to boredom. If we succumb, faulty widgets get past us. And we know that people who are more susceptible to being bored also have a greater tendency to do worse on sustained attention tasks. And when doing such tasks they report higher levels of in-the-moment boredom than do people who experience the state less frequently.[6] The difficulty in focusing attention experienced by people susceptible to boredom may give them more reason to seek refuge in distracting thoughts. Further, people who are susceptible to boredom are simply more likely to walk away and give up when asked to sustain their attention on a task that makes them feel bored.[7] Some of us are able to dig deep and stay focused on a boring task, but for others this is simply not possible.

When we're trapped by a boring task or circumstance, we quickly become restless. Sitting in that meeting at work outlining the new regulations for sending and receiving emails, delivered at a level six-year-olds could understand, most of us start to shuffle in our seats, leaning back and forward, stretching our arms behind our head, now folding them, drumming on the table, tapping our feet, and just generally feeling uncomfortable. You feel like you just can't take it anymore. You have to shake off the lethargy and hit the reset button hard. In moments like this, it can almost feel as if boredom threatens to strangle existence, making us desperate to feel *something* to confirm that we are alive. Such desperation can lead us down dangerous and unhealthy paths.

A stark example of this comes from a recent study in which people were asked to sit in an empty room and entertain themselves with only their thoughts.[8] For some, things did not go so well. Some reported difficulty concentrating, increased mind wandering, and a distinct lack of enjoyment.[9] When researchers gave people the opportunity to shock themselves rather than sit quietly, many chose the electric shock at least once, presumably in an attempt to alleviate the tedium. And they knew what they were getting into. Before the experiment began, they had felt the electrical shocks and had even said they would pay money to avoid having to experience them again. Yet all that went out the window when they were bored, many willingly reaching for the buzzer. Perhaps people in this study were simply curious about the electric shocks, as opposed to trying to use them to alleviate boredom. But before the boredom began, they all indicated that they would prefer to avoid the shocks. Nevertheless, two-thirds of the men and a quarter of the women chose to give themselves at least one electric shock. One man shocked himself 190 times in a fifteen-minute period! Curiosity should be satisfied more easily than that.

In follow-up research, people watched either a boring, repetitive movie several times over the course of an hour or a less boring movie for the same period of time.[10] When watching the boring movie, people shocked themselves far more often than they did when watching the interesting movie. This more controlled study clearly shows that it is something about being bored that is critical. And boredom may be a particularly potent trigger.

Chantal Nederkoorn and colleagues from Maastricht University in the Netherlands found that people self-administered more electric shocks when they were induced into a state of boredom than when they were induced to feel sadness.[11] Much like the bored mink in Chapter 1 that flirts with danger and seeks out aversive stimuli, it seems that boredom is so uncomfortable that people prefer painful physical sensations to the curse of monotony as a way to avoid the distress of boredom. The proverbial desire to poke your eyes out just to avoid the stultifying tedium!

Clearly, these experiments set up absurd situations that are hardly reflective of our daily lives. Importantly, none of these studies offered people options for alleviating boredom other than electric shocks. So we can't know what people would have done had they been free to choose any response to being bored. Nevertheless, some people do sometimes scratch, cut, burn, or hit themselves, in a manner similar to self-administering electrical shocks, when they are bored in everyday life. Such behaviors are referred to as "non-suicidal self-injury" (NSSI). Individuals engage in NSSI behaviors not out of any desire to end their lives but rather to use physical pain to help them feel better, to calm their distressed minds.[12] Self-inflicted pain might distract from emotional pain, create an escape from a difficult situation, and elicit help from others. Boredom, it appears, is so uncomfortable that it opens the door to such self-harming behavior.[13]

Beyond self-harm, a variety of other escape strategies are possible when confronted with boredom. Mind-altering substances, for example, offer escape and the promise of something better. Boredom is a disdain for the present moment, exaggerated when we feel powerless to change the situation. Substance use can offer a way to escape the present moment while it drags on without us.

When asked, many people who use harmful substances, such as cigarettes, alcohol, or drugs, report they do so out of boredom. These same people also report being bored more often than people who do not use harmful substances.[14] One survey conducted in 2003 found that 17 percent of teens report they are often bored, and those same teens are up to 50 percent more likely to report smoking, drinking to excess, and using drugs compared to teens who do not often feel bored.[15] A few studies have tracked levels of boredom and substance use in teens over time and confirm that boredom *precedes* the use of drugs, cigarettes, and alcohol.[16]

In one study of teens in both the United States and South Africa by Erin Sharpe and colleagues, a small increase in self-reported boredom was associated with a 14 percent increase in the likelihood of using alcohol. Boredom increased the likelihood of substance use 23 percent for cigarettes and 36 percent for marijuana.[17] Similarly, research with adults shows that recent episodes of boredom over the prior two weeks strongly predict alcohol use.[18] All of this research suggests that the state of boredom is a precursor of drug and alcohol use. There are hints that this link is even stronger in people who experience the state of boredom more frequently, suggesting that being prone to boredom is a risk factor for using substances to escape the feeling.[19] Boredom doesn't directly make us drink or do drugs. When we fail to exhibit any adaptive response, drugs and alcohol fill the gap by altering our mental state and ultimately numbing the negative feeling of boredom we want to escape.

Like substance abuse, problem gambling has been blamed on boredom. On the surface this appears to fit the same story told about harmful substance use. We get bored and the state is uncomfortable, pushing us to seek outlets to resolve the feeling. Gambling, particularly the engaging play of slot machines, fits the bill. We know that people who gamble often say they do so because they are bored, and there are a handful of studies showing that a stronger tendency toward boredom does predict the likelihood of having gambling problems.[20] Despite the common conviction that boredom leads to gambling, the research in support of this is the thinnest of all. It seems plausible that, like other vices, gambling may serve to relieve boredom, but we simply can't say so with confidence at this point.[21]

Like substance abuse and gambling, problem eating is on the list of ills boredom begets. When people are asked if they eat more when they are bored the answer is a resounding and consistent 'Yes!'[22] And when asked to describe the four actions they would most likely take if feeling a range of emotions, "eat" was much more frequently associated with boredom than with sadness or anxiety. So, based on self-reports, boredom appears to be a distinct trigger for eating.[23]

Andrew Moynihan and colleagues from Ireland and the United Kingdom asked people to keep a daily diary for one week documenting how bored they were each day and what they ate. They found that the more bored a person was on a given day, the more fat, carbohydrates, protein, and overall calories they consumed. This link between daily feelings of boredom and food intake occurred regardless of any other negative feelings like stress and was also independent of body-mass index and the general trait tendency to be bored. So, while other research shows that people who often feel bored report unhealthy eating, this study suggests there is something specific about feeling bored right now that is linked to higher levels of food consumption.[24] The same tendency was demonstrated in a lab study in which people were first made to feel bored by doing a dull puzzle. Then they were asked whether or not they felt like snacking or eating something healthy.[25] Bored people, who had a tendency to focus on their thoughts and feelings, reported a stronger desire to snack compared to another group who had not been bored.

So when we're bored we seem to eat as a way to occupy ourselves and stave off boredom. And we likely do so with less-than-healthy choices. One study showed that when people watched a monotonous, boring movie for one hour they ate the equivalent of 100 calories worth of M&Ms.[26] When people watched an interesting movie they ate half as many M&Ms. Notably, this tendency to eat when bored is not seen for other negative feelings. Boredom, it turns out, can't be equated with sadness (induced by having people watch sad movies). It was only boredom that led people to snack on unhealthy foods.[27]

Not only does boredom push us toward unhealthy eating choices, it may lead us to ignore the physiological signs telling us that we are satiated and in no need of more food. In what is now a classic study, Edward Abramson and Shawn Stinson from California State University, Chico, brought people into the lab and had them eat as many roast beef sandwiches as they wanted until full.[28] One-half of them were then asked to write the letters "cd" repeatedly, which was intended to bore them. Meanwhile, the other half were asked to write brief stories in response to interesting pictures. Both groups could snack as they pleased. People who were bored ate more than the others even after having feasted on roast beef. Sound familiar? Recall in Chapter 1 the mink housed in unenriched cages who ate more snacks than their brethren housed in enriched cages—this despite having been well fed beforehand.[29]

Boredom may even play a role in obesity. We have known for some time that genetic factors are linked to obesity. But what are the mechanisms behind this link? Clearly, such questions are extraordinarily complex, but one recent study raises an intriguing possibility that implicates boredom as a linchpin. Richard Gill and colleagues at Columbia University found that eating out of boredom and an inability to resist temptations partially explained the link between genetics and obesity.[30] Other negative feelings, like anxiety, were found to be unrelated to obesity.

We may turn to food when we're bored for many reasons. Perhaps in the throes of boredom we feel as though we're lacking energy. Turning to snack foods high in sugar instead of healthy snacks seems to support that idea. Alternatively, we may turn to food merely as a way to distract ourselves from boredom. The mere act of putting something in our mouths may delude us into thinking that we are well occupied. Finally, boredom may make us more vulnerable to impulses; that is, we see something enticing and unthinkingly put it in our mouth—sort of like going on autopilot. Like other consumptive tendencies associated with boredom, eating likely reflects a maladaptive, perhaps impulsive, response to escaping the feeling.

A number of studies have confirmed a link between the tendency to be bored and the tendency to act impulsively. We've mentioned elsewhere that boredom susceptibility is strongly related to sensation seeking—not the kind related to information gathering but the kind that seeks some new feeling to replace the drudgery of boredom. In this light, overeating, gambling, or consuming drugs and alcohol could be cast as impulsive attempts to alleviate boredom.[31]

This tendency to choose a short-term fix can make us act against our own self-interest. In one study, people were made to endure a short wait or to transcribe technical references about concrete. Both instances led to boredom, and in both cases people gave up the chance at a future bonus to avoid having to do more of the boring tasks. That is, the more bored people felt the more they chose to take the quick, easy reward (and escape from boredom), even though it was not financially in their best interest to do so.[32] In another study, people were kept waiting for five minutes before playing a dice game. The wait made them bored and resulted in risky decision making during the game.[33]

It's unclear whether people make risky choices when bored as a way to relieve the distress of boredom or merely that they fail to apply the brakes to their behavior. The consistent association between a tendency to boredom and low levels of self-control would suggest that people who get bored a lot fail to appropriately control their impulses. But these people may also seek thrills and excitement to replace the dullness that is boredom. In one study, those who were more prone to boredom reported making risky driving choices around train crossings, trying to "beat" the train across the tracks more often than people who were not prone to boredom.

In a more prosaic test of this, researchers had people to try to override an automatic tendency to look at something flashed on one side of a computer screen and instead look in the opposite direction. This is a classic experimental ruse used in cognitive psychology—show something shiny in one place and ask people to do their best to ignore it. It is a good measure of impulse control. This particular study examined smokers who were trying to quit, something that clearly requires the individual to control the impulse to smoke. It turns out that refraining from looking at the shiny object was difficult for smokers who had recently quit—as recently as a few hours before the test. This failure to avoid the allure of things in the environment (we could think of the shiny object as being similar to cues for smoking), is one of the things that makes it so difficult to quit. As it relates to boredom, the key finding was that smokers who reported experiencing higher boredom levels in their day-to-day life showed the greatest difficulty in resisting looking at the shiny object.[34]

Impulsive and maladaptive responses can be seen as attempts to address the unpleasant nature of the in-the-moment feeling of boredom. Digging deeper, we can see the particular pain of boredom—we feel paralyzed and helpless. We are unable to chart a course of action and follow through on it. We become inert. In short, we feel inconsequential, superfluous.

To be bored is to fail to be the author of our own lives. In this sense, boredom becomes an affront to our personhood. We are debased, an affront that some may not be able to endure without acting out. In fact, boredom may be particularly threatening because it saps the capacity for self-determination without indicating a clear oppressive force responsible for our plight. In this way of thinking, boredom becomes "rage spread thin."[35] If we avoid slipping into despair, one option is to lash out at the world. In one study, after 12- and 13-year-olds were asked to sit in silence for only seven minutes, these teens (and tweens) reported feeling more narcissistic and aggressive impulses compared to students who were not bored.[36]

We also know that people who often feel bored tend to be narcissistic.[37] Beneath the surface they are sensitive to any indication that they may be inferior or powerless. At the same time, boredom-prone people also experience higher levels of anger, aggression, and hostility.[38] Perhaps those prone to boredom exhibit elevated levels of anger, aggression, and hostility because they often struggle with boredom and the narcissistic slight to their ego that it brings. They may turn to aggression as a way to inflate their sense of self. A similar process, although likely not the root cause of the condition, might even be at play for psychopaths. It has been known for some time that psychopaths are particularly prone to boredom,[39] perhaps due to a temperamental need for excitement and thrill. And because psychopaths often exhibit strong narcissistic traits they are easily provoked to rage by circumstances, such as boredom, that challenge the grandiose image they feel they must defend at any cost.

In moments of extreme boredom the need to reaffirm of our sense of agency may become more important than the virtue of any particular action. We are driven to act simply to prove we can. Vandalism, cast in this light, is not merely senseless destruction but a response to challenged agency at least sometimes driven by boredom.[40] Our point here is that how we respond to threats to our agency will determine whether boredom has positive or negative consequences. We have the capacity and opportunity to destroy or create. Either will prove we are powerful and set the world right again in our eyes. But of course these options have very different consequences, both for us and for others.

Researchers who followed a group of 10-year-olds for five years found that a tendency to be bored at one point in time increased the likelihood of delinquency in the near future. For the whole sample, this was a one-way street: boredom proneness led to an increase in subsequent delinquency but not the other way around. But for a smaller group who were disinhibited and had a high need for excitement, delinquency at Time 1 actually predicted *lower* levels of boredom proneness at Time 2. For sensation-seeking youth, delinquency may have relieved, or buffered against, future boredom.[41]

In an experimental study, researchers made a group of university students bored and then asked them to read scenarios describing a member of their own cultural group being attacked by a member of another cultural group (or vice versa).[42] When people were bored they assigned more lenient jail terms to offenders from their own cultural group and more severe jail terms to offenders from another cultural group.

In the most severe cases, when coupled with an extreme lack of empathy, boredom can be highly dangerous. "We were bored and didn't have anything to do, so we decided to kill somebody."[43] This was the reason offered by one of three teens who senselessly murdered a young man who was out jogging. There are numerous examples of murderers claiming they killed out of boredom. But surely boredom didn't directly cause the teens to resort to murder. Presumably, they had other avenues to achieve excitement and reassert control over their lives when boredom overcame them.

More recently, a German nurse was suspected of killing at least ninety-seven people.[44] He claimed he killed to alleviate boredom. He so enjoyed the rush and acclaim of being able to resuscitate patients that he deliberately injected them with dangerous drugs so that he could then revive them—all to show off his skill and ward off boredom. The euphoria he experienced when he brought a patient back from the brink of death and the dejection when he failed were both elements of an out-of-control exercise in power. Clearly, boredom by itself isn't enough to push the average person to act so abhorrently. Boredom signals we are unengaged and not the masters of our lives. This feeling may lead some to lash out and seek extreme forms of power.

Delinquency, anger, hostility, and violence may do double or even triple duty as a response to boredom. Such behavior may bolster our sense of potency, which is threatened when we're bored. We might not be "in control" in the colloquial sense when we lash out, but the effects of our actions are obvious—we caused the destruction. Also, anger and violence may invigorate and stir us from the heavy lethargy often associated with boredom. Boredom and monotony are often felt as low arousal states. If we want to redress that slump in energy, then aggression certainly gets the blood pumping. Finally, hostility may reduce boredom in the short term by making the world seem more meaningful.

The aggression and hostility born of boredom is sometimes targeted. As we discussed earlier, bored people gave less severe jail sentences to members of their own cultural group and more severe sentences to people in cultural groups different from their own. Jingoism—an extreme form of nationalism—has been shown to occur when one's sense of meaning and purpose in life is threatened.[45] Harboring negative views of people we do not identify with and positive views of our own tribe makes the world simpler and easier to understand. Acting aggressively toward outsiders creates a confident, powerful feeling, reduces anxiety, and provides the illusion that the world is simpler, more understandable, and more stable—even if that world order is morally reprehensible! We explore this relation between boredom and meaning in more detail in Chapter 7. The point here is that at least some forms of aggression could be viewed as attempts to redress the lack of meaning that is associated with boredom.

What we need and ultimately want is to feel meaningfully connected to the world. We do best when actively connected: using our cognitive abilities, giving expression to our ideas, and mastering our surroundings all serve to keep the system humming. In contrast, boredom represents a state of being disconnected, and this makes us vulnerable to problems that meaningful engagement would otherwise keep at bay.[46] The well-established link between depression and boredom proneness, for example, may reflect such an inward consequence of disengagement.

People who often feel bored are also more likely to struggle with depression.[47] At first glance, depression and boredom may seem to share so many characteristics as to be indistinguishable. But they are different. Depression is defined by sadness and an inability to feel pleasure. It is associated with negative self-evaluation and a tendency to focus on negative life events. In contrast, boredom is defined by the conundrum of wanting to be engaged but being unable to satisfy that want, a feeling that time is dragging on, and a difficulty in concentrating. And, in contrast to depression, boredom is associated with negative evaluations of the world outside of ourselves, a lack of emotional awareness, and a combination of restlessness and lethargy. Clearly, boredom and sadness are different beasts.[48]

So what is it that explains the relationship between boredom proneness and depression? Some research suggests that the tendency to be bored and depression are part of a vicious cycle that plays out over time. Michael Spaeth and colleagues[49] tracked more than 700 adolescents for five years, each year asking them about depression and boredom proneness. Their results showed that depression and boredom proneness intensify each other from one year to the next with no clear indication of which came first. However, in our own lab we failed to find any evidence that depression predicts levels of boredom proneness over an eight-week period. We also found that when people had their mood changed by recalling a happy or sad memory it had no impact on their levels of state boredom.[50] One of the authors of this study asked individuals hospitalized because of chronic debilitating depression about their experiences of boredom. The pattern that emerged from their stories was clear. They reported being *afraid* of boredom. For them, boredom was an early warning sign that depression was on its way. When bored, they were thrown out of engagement with the world. They stopped feeling passionate commitment to activities and people, and when this happened, they turned inward and started to ruminate. They started to think about themselves in negative ways and, in time, these thought patterns spiraled into a full-blown depressive episode. So, taken as a whole, there is stronger evidence to support the idea that boredom leads to depression as opposed to the other way around. It may be that boredom leaves a person vulnerable to rumination and negative self-focus as they continue to struggle to engage with the world, eventually succumbing to despair in response to a continued failure to engage.

Another possibility is that the tendency to experience boredom and depression are related because they are both caused by some third factor. This notion is consistent with another finding from the work of Spaeth and colleagues. They found that depression and boredom have similar trajectories over larger time scales, suggesting they may share a common developmental factor and be part of a larger syndrome that unfolds over time. That common factor might be feeling as if life lacks meaning and purpose. It could be that an inability to establish and follow through on valued life projects is a driving force behind both depression and boredom proneness.[51] Our own work is consistent with this possibility. Decreases in life meaning appear to cause increases in the state of boredom.[52] So, while it is possible that depression and boredom proneness follow from a lack of life meaning, more work is needed to confirm this possibility.

While we may not know why or how boredom proneness and depression are related, it is unfortunately clear that boredom is typically ignored when we assess and treat mental health. Yet, boredom may play a key role in a number of mental illnesses.[53] In one study, boredom proneness was shown to have a significant negative impact on the quality of life for cancer patients—over and above any negative impact of depression. Also, when those same patients were treated with antidepressant medications their mood improved, but overt boredom proneness did not.[54] These striking findings add further weight to the idea that boredom proneness and depression are indeed different problems and that boredom requires its own intervention. It is not enough to simply treat complaints of boredom as just "related" to depression.

Although boredom does not directly cause problems all by itself, the case against it seems pretty strong. At times it goads us into inflicting damage on our self and others, and at other times it renders us more impulsive, susceptible to influence, and vulnerable to the strain of being disengaged. Clearly, responding adaptively to boredom is no easy task. Nevertheless, there is at least one thing researchers have suggested boredom might be good for—as a spur to creativity.[55]

Mike Bloomfield of the Paul Butterfield Blues Band, an eventual inductee into the Rock and Roll Hall of Fame, was stunned by the brilliance of a guitarist playing a right-handed Fender left-handed and strung upside down (upside down for a leftie anyway). He cornered Jimi Hendrix after the show and asked him, "Where have you been hiding man?" Hendrix replied, "I been playin' the chitlin circuit and I got bored shitless. I didn't hear any guitar players doing anything new and I was bored out of my mind."[56] Hendrix's assertion suggests that being bored can lead to something positive—creativity.

But what do the data say? Actual demonstrations that boredom leads to creative flourishing are few and far between. One study suggests that our capacity for writing creative essays drops substantially when bored. Unfortunately, this study had no comparison group. People were not induced to feel more or less bored.[57] In another study, researchers induced people into various states, including boredom. Boredom, when grouped with elation, was associated with increased creativity.[58] But the researchers did not separate these two very different psychological states. Finally, Sandi Mann and Rebekah Cadman from the University of Central Lancashire made some people bored by having them write out or read numbers from a telephone book. Afterward, they asked them to think of as many possible uses for a polystyrene cup as they could—a classic task used to examine creativity.[59] Those people who were bored *and* who reported daydreaming came up with more creative uses for the cup than did those who had not been hit over the head with the telephone book (metaphorically speaking). As with the problem of distinguishing boredom and elation, it is impossible from this study to know whether daydreaming or boredom was the key to creativity. Ultimately, when we're daydreaming we're no longer bored; instead we are engaged in internal reverie.

So there is no clear evidence that being bored leads to creativity. Just as boredom doesn't make you a killer, it also fails to turn you into a creative genius. And that is the point. Boredom is a negative feeling that we want to expunge. But by itself is does not make us act in either good or bad ways. It makes sense to say that the *capacity* to be bored can push us toward creativity and innovation. Hendrix, not content with the status quo, forever changed the world of guitar playing. We seriously doubt, however, that he was in the throes of boredom while creating *Foxy Lady*.

Boredom is the unfulfilled desire to be mentally occupied. It's an uncomfortable feeling, motivating action. The action we take is up to us. We can respond to the signal by turning to drugs and alcohol, or by lashing out aggressively at the world. Or, like Hendrix, we can pick up a guitar and make something wonderful and, in so doing, no longer be bored. While creating his masterpieces Hendrix was not bored; he was deeply engaged and connected with his guitar and the music. Such connection is, in fact, the antithesis of boredom. When bored it's as if we lose our connection to, or become isolated from, the world around us. As the sociologist Peter Conrad put it, "Boringness isn't out there; it is between there and us . . . [boredom is] alienation from the moment."[60]

CHAPTER 6

BOREDOM AT THE EXTREMES

• • •

The walls are bland and the building labyrinthine. So much so you suspect merely getting there is a test.

Eventually, you find room P4040 and knock, more than just a little nervous. Your roommate lasted three days, and you're not sure you'll fare so well. The door is answered by a lab tech, who offers you a seat and begins to explain.

You'll be placed in a small room. You can leave only for meals and bathroom breaks. The room is brightly lit with a uniform white light. An air conditioner is constantly running, filling the small space with a steady hum. There is a cot and nothing else. You will be required to wear opaque glasses. You can open your eyes, but all you will see is uniform, blank, unending whiteness. To top it off, you must wear an odd set of gloves that constrict movement and limit the sensation of touch. You won't even be able to scratch yourself.

"We want to know how humans deal with monotony. Do you have any questions?"

You shake your head nervously, take a quick bathroom break, and briefly consider making a run for it before returning to start the experiment.

> Once set up, your thoughts start to settle. *This is not so bad,* you think. But it isn't long before your mind begins to wander, from your course load to a party you want to go to on the weekend. Eventually, you sink into nostalgic reverie, recalling personal stories from your life, like the time your older brother taught you to ride a bicycle. You try to reimagine them in great detail. Your body has been restlessly shifting around on the cot, unable to settle into one position. At some point you wonder, *How long has it been?* Later still, you find it hard to maintain a single thread of thought. Nothing connects. Before you've finished one thought, another intrudes.
>
> That's when you see it. A shadow at first, morphing quickly into the figure of a man. Tall, a long overcoat perhaps, no discernable features, but a quiet, lurking menace. *Is this real? Is this part of the experiment?* Your breathing escalates. Time to call it quits.

• • •

The sensory deprivation experiments conducted by Donald Hebb, Woodburn Heron, and their colleagues at McGill University, first published in the 1950s, had the stated aim of exploring behavioral responses to monotony. But these experiments did far more than subject people to monotony; they deprived them of any sensations. Beyond monotony, isolation—environmental and psychological—looms large as a component of boredom.[1]

As we outlined in Chapter 2, monotony is indeed a prime driver of boredom. But for Hebb and Heron, exploring monotony for monotony's sake was a smoke screen. The truth behind their odd experiments was something different altogether. Canadian, British, and US intelligence services wanted to know why prisoners of war returning from the Korean theater were now espousing Communist ideologies. How had the Koreans achieved such brainwashing? At a meeting Hebb had serendipitously been invited to, he suggested that perceptual isolation may achieve that end and proposed experiments to test the notion.[2] The results had broader implications for how humans cope with a sparse sensory milieu.

Hebb and Heron's experiments showed, quite starkly, our deep need to interact with our environment in autonomous and self-determined ways. Avoiding boredom is not simply about avoiding monotony. Nor is it simply about finding meaning, which we delve into in the next chapter. Avoiding boredom is also about finding ways to interact with the world and others in it. In their experiments, people spent days on end in the specially designed room, presumably motivated to stay by the monetary pay off. Regardless of the duration of stay, most people drifted into sleep early on. Without things to do, sleep was a viable option. Upon waking to find their world unchanged, the most common experience was one of restlessness.

People also commonly reported being unable to form clear, concise trains of thought. Without sensory stimulation to guide them, thoughts seemed to drift aimlessly and were unconnected to one another. It was hard to focus attention in order to corral thoughts into some meaningful, coherent whole. We need to feel connected with the world, to interact with things tangible to the senses. Without that connection, our cognitive systems do not run normally. As time wore on, hallucinations of a wide variety were common, something also reported by aviators on long, monotonous flights, who have reported large spiders appearing on the windshield! Initially,

Hebb and Heron employed their methods to see whether sensory deprivation, combined with "propaganda" about the supernatural, would prompt people to shift their beliefs. Could this be an explanation for the Communist brainwashing of prisoners of war? After the study people were indeed more willing to believe in the supernatural. Whether such a change in beliefs lasted very long or would also work on ideological convictions is less clear.

Sensory deprivation is an extreme situation in which we are disconnected from the world. Under isolation our cognitive system falters, and we are not even able to seek refuge in focused, coherent daydreaming. Unable to become mentally engaged, sensory deprivation leaves us terribly bored. Beyond these studies we find there are other circumstances, no less extreme, in which isolation and boredom are unhappy bedfellows. What this highlights is not only that boredom is a call to action, but that a basic, fundamental human need is to be connected to others and an actor in our own stories. Without the fulfillment of those needs, we feel superfluous and ultimately bored.

Extreme Environments

Humans are born explorers. Who hasn't felt, upon seeing the majesty of the Rocky Mountains, the desire to climb them? Or wondered, when looking out at a vast lake or ocean, just what might be on the other side? Whether this urge to explore is due to an innate drive to discover new frontiers, a restless kind of curiosity to see for ourselves what lies beyond the next mountain range or oceanic horizon, or simply a product of our desire for fame and fortune, we have consistently pushed ourselves to the ends of Earth and beyond. The world beyond the mountains or beneath the oceans and the frontiers of our galaxy and universe call out to us.[3] At times this has led us to the exploration of some of the most barren and inhospitable places on Earth—the polar regions of Antarctica and the Arctic Circle. Beyond the formidable physical challenges posed by these extreme environments lie the challenges of an isolated life, the monotony of sensory inputs and the rigidity of the routines needed to survive. While mapmakers may have imagined monsters on the blank spaces of ancient maps, they failed to anticipate the demons of isolation and boredom that would haunt those who explored those realms.

Studying the psychological consequences of exploring and living in extreme environs, sometimes referred to as Isolated Confined Environments (ICE), is difficult.[4] The missions are few and have goals typically not focused on psychological research. Team composition on such expeditions is highly variable, and the number of team members tends to be small, making it difficult to extrapolate findings to the general population. Inherently high risk, each expedition faces distinct challenges that in turn will influence the psychological experience of the team members. So what we learn from such expeditions must be considered cautiously.

The voyage of the Norwegian-built, Belgian-commissioned ship the *Belgica* in 1898 may have been our first hint that the isolation and confined nature of these extreme environments might lead to boredom. The ship was the first to overwinter in the Antarctic, and despite inadequate equipment and clothing and the ever-present threat of scurvy, it was monotony and the slow drag of time that impressed the expedition's American doctor, Frederick Cook: "We are imprisoned in an endless sea of ice, and find our horizon monotonous. We have told all the tales, real and imaginative, to which we are equal. Time weighs heavily upon us as the darkness slowly advances."[5]

One crew member got so sick of the isolation that he left the icebound ship and claimed he would walk home to Belgium! In this story, we see many of the factors that are fertile ground for boredom. First, there is the monotony arising from isolation: in both the environment itself—"an endless sea of ice"—and in the activities available for pursuit—the tales having all been told. Second, time dragged on, passing without any discernible change in their environment. And though Cook doesn't mention it in this quotation, they faced an almost complete lack of autonomy: the crew were at the mercy of the elements, with individual actions unlikely to have any influence on the outcome. These factors have been confirmed in analyses of team members' complaints of psychological challenges on more modern expeditions in extreme environs—from space to polar exploration. Across these very different environments, the most common challenges mentioned include a diminished sense of autonomy or control over the situation, interpersonal conflicts, and a restlessness driven by boredom.[6]

Some suggest that different stages of ICE missions have distinct psychological signatures. The earliest stage is best characterized by anxiety, followed in the middle stages by depressed mood and boredom, with the final stage manifesting in immature behaviors and a sense of euphoria now that the end is in sight.[7] Common across studies is the notion that when boredom does arise it is best described as a sense of restlessness—a desire to engage in something that is thwarted by the demands of the situation. Isolation from family and normal social interactions and physical confinement in harsh and inhospitable environments compound the issue.

Personality differences also play a role. Those more agreeable and more open to new experiences tend to do better in isolated environments. Those high in neuroticism—a personality variable we know is common in those prone to experiencing boredom—don't fare quite so well.[8] Interestingly, a *lower* need for stimulation is also protective against boredom in extreme environments. This hints at the challenge of finding a good fit between team members and environment for such expeditions. If you're the kind of person who craves constant change and novelty, the endless Arctic tundra is probably not for you.

It should be noted that things are not all bad in isolated, extreme environments. There is a salutogenic effect if we are open to the experience, a sense of awe at the surrounding vastness and our place within that grand scale. This highlights the strong link between personality profiles and the capacity to cope with such environs—and to cope with boredom. Those who are more open to new experiences and who need less external stimulation likely experience far less boredom than those high in neuroticism and hungry for a constantly changing world. There is little research into an individual's capacity for experiencing awe.[9] Think of the feeling you have upon first entering an enormous Gothic cathedral, or standing on the rim of the Grand Canyon. Words may not suffice, but the feeling is one of awe. For those who effortlessly experience awe in a surrounding that is vast but ultimately uniform, boredom may pose less of a challenge. That is a good thing, because boredom in such extreme environments has potentially serious ramifications.

One study suggested that in space missions, attention and psychomotor functions showed some decrement in performance over time, whereas mental arithmetic and memory did not.[10] We know that boredom is strongly associated with failures of attention, so in terms of our understanding of the experience of boredom this finding is hardly shocking. But it does show that boredom in extreme environments could have serious consequences. When the requirement for vigilance is high, but the task is routine—the signature of monotonous, boring circumstances—boredom will be detrimental to success (Chapter 5).

Combat in war zones represents another extreme environment where, perhaps counterintuitively, boredom likely plays a role. The adage that war is best characterized as months of boredom punctuated by moments of sheer terror suggests that boredom arises at distinct stages of the experience (Figure 6.1).[11] The same could be said for peacekeeping missions.

A study examining psychological stressors in American peacekeepers operating in Yugoslavia in the 1990s suggested that five key psychological factors were at work; isolation, ambiguity, powerlessness, danger or threat, and boredom.[12] Just as for polar and space missions, isolation from family, a sense that one lacks autonomy, and boredom are prevalent in war zones, despite the stark differences in mission. In all such missions, boredom was more likely to arise mid-mission. Persisting into the later stages of deployment, boredom was associated with monotonous work routines, interpreted by soldiers as meaningless "busy work"—that is, doing something for the sake

Figure 6.1. This drawing, made by Sidney Gunter, a Canadian enlisted soldier in World War I, suggests that even in the midst of constant bombing, soldiers can feel bored.

of doing something. Making beds, folding uniforms, and polishing boots are not really necessary for combat readiness, but they are a good way to occupy time, though still boring. We need to be acting in and on the environment, but just any old action won't suffice. As we touched on briefly in Chapter 2, self-determination is critical. Activity must flow from our desires and occupy our minds to stave off boredom. Activity that is forced on us, or disconnected from our desires, will feel worthless and will not engage us. Being occupied with pointless activity, rather than something we find to be of value, is a sure way to bring on suffocating boredom.

Doing Time

Solitary confinement is another extreme form of isolation. First developed in the United States in the nineteenth century, one system of incarceration—the Pennsylvania system—advocated isolation for all prisoners. This was not merely isolation from society but from all other prisoners as well, presumably to facilitate a prisoner's need to practice penance (hence the term penitentiary).[13] A Supreme Court ruling in 1890 highlighted the raft of detrimental effects of such practices, yet it remains in use today under the guise of a multitude of euphemisms including "secure housing units," "administrative segregation," "quiet rooms," "communication management units," and many more.[14]

Not only does solitary confinement elicit boredom, it also makes it extremely challenging for the individual to self-regulate thoughts and actions, just as it did for the participants in Hebb and Heron's sensory deprivation experiments. This compounds the problem because boredom becomes all the more challenging when the capacity for self-control and self-regulation is diminished. A person confined to a cell for twenty-three hours a day will find it extremely difficult to string thoughts together and find meaning in lived experience. And clearly there are few to no outlets for self-determined action. The degree of sensory deprivation and the length of the stay correlate with negative psychological outcomes.[15] Many of those outcomes—increased anger, stress, and impaired concentration—are common bedfellows of boredom, which itself is prominent in those confined.[16]

Recent cases in Canada highlight the most extreme consequences of solitary confinement. Ashley Smith wrapped a cloth around her throat and died from choking while guards, instructed not to intervene so long as she was breathing, watched from outside her cell. She had been in solitary confinement for twenty-eight months. Adam Capay, a First Nations man charged with first-degree murder at age 19, was held in solitary confinement for more than four years—alone in a cell for twenty-three hours a day with a light constantly on. He was described as drifting in and out of consciousness and was often found bashing his head against a wall. These conditions are obviously not conducive to good mental health.[17]

Data from 2004 suggest that there were at that time 25,000 inmates held in solitary confinement in the United States. Current estimates place that number between 80,000 and 100,000.[18] With social interactions severely restricted and a dearth of materials such as books or writing implements to occupy the prisoner, life is completely determined by external structure. Each day's progress depends entirely on when meals are delivered and when outside time, however limited, is granted. In isolation, we are not the masters of our own destiny. In the extreme conditions of solitary confinement, any action we *can* perform is determined not by ourselves but by the routines imposed upon us from the outside. A stark description of this kind of dependence on external events can be found in Christopher Burney's account of his more than 500 days in solitary confinement as a prisoner of war during World War II. Burney structured his day by meal delivery times and the shadows cast on the stone walls from the little sunlight available. He pushed himself to save some of the morning's meager rations for later consumption. When he failed in that exercise of self-restraint, his mood plummeted. External structure was critical, and failures of self-control were devastating. Burney's confinement shone a light on the flip side of this need for routine and structure—a need for variety. As he put it, "Variety is not the spice of life, it is the very stuff of it."[19]

But what of boredom for the incarcerated—not those confined to a cell with little to no social contact? Incarceration still isolates inmates from society, removes the prisoner from their family and social networks, and imposes strict routines with severe limits on autonomy. In an ethnographic study of incarcerated youth in Denmark, observations of young people in the prison setting suggested that boredom insinuated its way into every aspect of their lives.[20] Feeling that everyday events and tasks were completely devoid of meaning and that autonomy had been stripped from them, the youth reported boredom as the most prevalent psychological experience. In some instances, the Danish inmates in this study described taking actions against their externally imposed routine, engaging in various forms of striking. Such attempts to regain some semblance of control over their heavily constrained world represent a direct response to being bored.

Tragically, many of the inmates reported that it was boredom and the desire to seek stimulation that had landed them in trouble in the first place. As we have outlined previously, high levels of boredom proneness are associated with increased risk taking and thrill seeking. In some instances, this can lead to criminal behavior.[21] That boredom is then a part of prison life is also prominently reported in adult prisoners in the United Kingdom. Here too, the inflexible, externally imposed structures of daily life are partly to blame.[22] With no control over what happens to them, with all rules and consequences imposed on them, told what they can and cannot do, and with little sense that what they do has any broader meaning, all things are deemed to be merely "killing time." Indeed, the colloquial description of incarceration, "doing time" is apt; there is nothing else to be done but to mark the time until release. Without specified goals to work toward or meaningful actions to engage in, time drags and boredom dominates. Time itself is not the demon here. Rather, it is the fact that in isolation we are robbed of our agency—our capacity to influence even our day-to-day routines, to the extent that the only thing that matters is time.

A Tale of Two Astronauts

As we've seen throughout this chapter, boredom has been shown to arise in solitary environments, from polar and space expeditions to incarceration. Yet, it need not be an inevitable consequence of such environs. The individual's reaction to isolation is critical. Brief moments of boredom are neither good nor bad for us. It is our response to the boredom signal that determines the outcome. What we have emphasized in this chapter is that isolation of various kinds constrains our options for responding. And when that happens, boredom becomes far more challenging to address. The experiences of two astronauts—well, one astronaut and one cosmonaut—highlight the importance of how we respond to isolation and the boredom it can engender.

Valentin Lebedev, a Russian cosmonaut, wrote a diary of his 211 days spent in space in the early 1980s. At the time this was a record stay. Lebedev's diary of the mission launched in 1982 is replete with anxieties and challenges in response to the isolation of space. Lebedev is strikingly honest in his appraisal of the challenges he and his crewmate faced, often reflecting on his own concerns and shortcomings. As mentioned earlier, missions of this kind are often marked by anxiety in the early stages, and Lebedev is vocal about his own experiences. He reports being concerned that the mission will not succeed and claims he is always "on edge." Lebedev spoke of time dragging; only a week into his mission he writes that the "drab routine has begun." Like the incarcerated youth in Denmark and peacekeeping soldiers in Yugoslavia, Lebedev reports doing what sounds like "busy work"—"Ground Control gave us a whole bunch of small things to do, which although important, were tedious" And later, "FCC [the flight control center] talked about petty things"[23]

His experience is a study in contrast to that of Canadian astronaut Chris Hadfield, who found fame through the Twitterverse while acting as commander of the International Space Station in early 2013. Hadfield experienced many of the same physical and practical challenges Lebedev did, but reported none of the anxieties.[24] In describing his time on the ISS, Hadfield also notes he had little challenge in finding purpose and a sense of meaning in even the simplest of activities. For him, even fixing the plumbing on a toilet in the space station was meaningful and valued. Hadfield maintained a capacity to find purpose and challenge in even the most mundane things, a skill that perhaps predated his time as an astronaut.

Hadfield claims to never be bored. He reported finding great satisfaction plowing the fields during his childhood on a farm in Southern Ontario. Progress was obvious and satisfying. In contrast, harrowing—the task of breaking up and smoothing out soil—was less satisfying to him. Behind your tractor was the same brown expanse you saw ahead of you, making it impossible to get a sense of progress and achievement. This highlights the fact that committing to an action is not enough to fend off boredom—we need to also see and value the consequences of our actions. But rather than letting harrowing sink him into a state of boredom, Hadfield challenged himself to see how long he could hold his breath while doing the mundane task. "You use maybe 30% of your brain on the primary task and with the rest you can dedicate to something else."[25]

Clearly, the two astronauts responded to monotony in very different ways. Hadfield found ways to fully occupy himself, whereas Lebedev unsuccessfully sought distraction from the mundane. Upon receiving a video camera from a supply ship Lebedev states: "Soon, we'll put the video recorder together and it won't be so boring here." For Lebedev, feelings of isolation—the underlying theme of this chapter—were common. He lamented the distance and time away from family, remarking that "everything is down on Earth."[26] Perhaps this thought reflected a feeling of disconnection from meaningful social interactions.[27] He was clearly as accomplished an astronaut as Hadfield. His unique and honest insights likely influenced the way in which missions are now deployed in hugely positive ways. And to be fair to Lebedev, he commonly remarks on the majesty of his surroundings. Yet, intermixed with his sense of awe, he reports feelings of frustration, boredom, and ultimately depression.[28] After five months in space, Lebedev laments that "our interest in work is fading. I don't even want to look out a porthole anymore." Hadfield, in contrast, has said that in every quiet moment on the space station this is precisely what he would do.

Anxiety, isolation, and routine can take a toll. In addition, Lebedev frequently reports feelings of frustration with ground control centering on requests he felt were unimportant; obsessions with his physical health that he saw as irritating and a general lack of autonomy. This lack of autonomy is even stressed in relation to the instrumentation he relies on: "instead of being masters of the equipment, we are its slaves."[29]

The contrast between Lebedev's and Hadfield's descriptions of life in space speak to what we would call a person-environment fit. Lebedev is a pioneer in science and space travel, but he found aspects of isolation in space challenging. Hadfield was able to mitigate these challenges. It is not the isolation or the routine of a given experience or environment that is inherently boring or frustrating. It is our response to it that matters most. Underlying it all are the dual needs to express ourselves to our fullest capacity and to find ways of being mentally engaged. Isolated in an extreme environment, this is far harder to do. Prisoners in solitary confinement provide the most stark example of how isolation from normal engagement thwarts our ability to exercise our skills and express our desires. In this most extreme of circumstances there is little to nothing for the individual to engage in, and prolonged isolation of this kind has extremely negative consequences for mental health. Similarly, soldiers trained to engage in high-stakes activities don't respond well to being forced to do mundane "busy work" simply intended to occupy their time. In all these instances, what we *want* to do or what we *know we could do* and what the environment or current circumstances *allow* us to do don't match up. Boredom is one prominent outcome of such poor person-environment fits. And, in such moments, when we can't exercise our skills and express our desires, when we are forced to do things we don't want to do, when we can't stay engaged with what's on offer, an oppressive sense of meaningless slowly and menacingly starts to creep into our soul.

CHAPTER 7

THE SEARCH FOR MEANING

. . .

The talk seemed promising from its title. But fifteen minutes in, you fail to even recall what the title was; something to do with the bacterial ecology of the Thames. Not exactly your area of expertise, but close enough to make it worth your while. Or so you thought.

You had been alert and eager to begin with, the introduction engaging and meaningful enough in its own way. But somehow the point of the whole talk had unraveled; perhaps the speaker's monotonous drawl was to blame, or maybe the topic was a bit too advanced for you. Either way, now you find yourself shifting in your seat, slumping down one minute, then leaning forward, head cradled in your hands the next. You look around at the other audience members. Incomprehensibly, some seem to be enraptured with what you've now decided is pointless, bereft of any semblance of meaning for you. Others are also shifting restlessly, unable to maintain a decorous posture.

That's when you spot your colleague Dr. Galton sitting one row ahead, to your right. He's scribbling some notes, and you try discreetly to peer over his shoulder to discern what he could possibly be getting out of this long, tedious lecture.

Galton seems to be studiously observing the audience more than attending to the speaker. What could he be doing? You scan the audience, trying to see what he sees.

Later, when you run into Galton, you ask him what he was doing.

"Trying to gauge the audience's satisfaction with the presentation, my dear man. I didn't find the thing to be altogether very engaging, and my posture was a dead giveaway. I realized I was not alone. When attending deeply, rapt by the meticulous information presented, our colleagues sat upright and had minimal sway. But when they failed to adequately focus their faculties on the poor fellow schooling us on the precise constituents of that trail of mud we call the Thames, they began to fidget and sway."

Galton had chosen to spend his time eking meaning not from the topic of the talk itself but from something he habitually derived pleasure from—the measurement of human behavior.[1]

• • •

We live in an interpreted world: all that we see, smell, hear, taste, and touch is colored by the meaning we assign to it. We see patterns. We see purpose, value, and significance. Sensations are judged as good or bad. We expend resources, relinquish large amounts of personal time, and in some cases give up our lives, all in the quest for meaning. Firefighters put themselves in harm's way, terror groups commit numerous atrocities, and the über-wealthy donate large portions of their net value. We may come to different answers, but it seems impossible for any of us to refrain from giving meaning to our experience. Boredom is integral to this core need. It tells us we have lost meaning and can motivate us to find it again.[2]

Lars Svendsen from the University of Bergen asserts that we live "in a culture of boredom" reflective of a crisis of meaning in society more broadly. "Boredom," he suggests, "can be described metaphorically as a meaning withdrawal. Boredom can be understood as a discomfort, which communicates that the need for meaning is not being satisfied."[3]

Our motivation to seek and find meaning is perhaps most poignantly demonstrated by Victor Frankl's harrowing account of life in a Nazi concentration camp. What he calls a "will to meaning" is essential for enduring the most inhumane circumstances. Further, this will to meaning holds the key not just to survival but also to human flourishing. When meaning is absent, we are left with an inner void or emptiness, which he calls an "existential vacuum." According to Frankl this absence of meaning is at the root of much of human suffering and misery. Boredom, he suggests, is a central player—"the existential vacuum manifests itself mainly in a state of boredom."[4]

Philosophers, theologians, and writers have been wrestling with these themes for a long time. Reinhard Kuhn, for example, in his now classic book, *The Demon of Noontide: Ennui in Western Literature,* has insightfully traced the idea of boredom across history, exploring how it both reflects and shapes our reality.[5] In contrast, scientists are relatively late to the game. What does the relatively recent research tell us about the relation between our need to make meaning and the experience of boredom?

Meaning Lost

Wijnand van Tilburg, a social psychologist at King's College London, and his mentor Eric Igou at the University of Limerick are leaders in the experimental study of the relation between boredom and meaning. In a set of foundational studies, they asked people to recall a boring time as well as a time they felt sad, angry, and frustrated. In another study, they elicited boredom by having people copy out bibliographic entries about concrete. In each case, boredom was uniquely associated with feelings of meaninglessness.[6] For van Tilburg and Igou, a lack of meaning is the defining feature of boredom.

In our view, if a person is bored, they will indeed feel that what is happening is pointless and has no significance. This lack of significance is the result of not being engaged with a desired activity in the moment. If you had a job like Humphrey Potter's (Chapter 3), pushing one button at the right time and then another some time later, it is hard to imagine how you would find meaning in that monotony. However, we believe it is possible to feel that an activity we are engaged with—and therefore not bored by—can lack significance. So there is an asymmetry in the relation between boredom and a lack of situational meaning.

Researchers have also explored the link between the broader sense of life meaning and boredom. To illustrate the distinction between situational and life meaning, imagine those who generally feel as if their life has a great deal of meaning and purpose. They have a coherent set of values and beliefs that make sense of their experience and guide their actions. Yet, on Saturday morning as they wait in line at the grocery store, they feel that they are wasting time doing something pointless. We would say that they possess a great deal of life meaning but find standing in line to lack situational meaning. The key difference is the reference point for the feeling of meaninglessness. In one case, the reference point is one's life as a whole, and in the other, it is tied to a particular situation. To be clear, the exact boundary is not always easy to find. One might bleed into the other, sometimes in imperceptible ways and at other times more obviously. Think of doing something menial in your job, such as filling out a record of the day's activities for billing purposes, in fifteen-minute increments. You like your job and find it contributes to your sense of life meaning, but this particular situation may make you doubt yourself. Nevertheless, we contend the distinction is important for understanding the relation between boredom and meaning.

People who feel that their life lacks meaning also frequently feel bored.[7] This is true in younger and older adults and can be seen using both direct and indirect ways to define and measure meaning in life. Individuals who are politically engaged report less tendency to feel bored, and people who report achieving their goals, political or otherwise, also report less boredom.[8] Boredom and meaning are tied together regardless of how an individual succeeds or fails to make meaning.

In our lab we found that people's beliefs about how meaningful their life is predicts their likelihood of feeling bored in the future.[9] This finding is important because it shows that boredom and meaning are related *over time* and points to the possibility that beliefs about the meaningfulness of life may actually cause boredom. Our findings are backed up by other research. For example, on the basis of clinical case studies researchers have noted that the people who often felt bored were distinguished by the fact they had all failed to find a large-scale project that gave their lives meaning.[10] We don't all need to build a multimillion-dollar computer software empire that enables us to found a philanthropic organization to feel we have meaning in our lives, but some broader, long-term goals are helpful.

Richard Bargdill of Duquesne University argues that having compromised an important life project is the key source of chronic boredom.[11] He conducted in-depth interviews with people who often felt bored and discovered that they had never come to terms with having given up on important life projects. They claimed others and life circumstances, such as poor teachers or illness, prevented them from pursuing their dreams and goals. But under the surface they were also angry with themselves for giving up and not following through. They were not able to fully engage with life because they were not doing what they really wanted to do. Moreover, they became pessimistic about the possibility of future life satisfaction. Eventually, as time went on, they became more passive, defensive, and withdrawn.

Studies like this provide a tantalizing hint that a lack of life meaning *causes* boredom. But they are correlational, and as all budding young scientists are taught, correlation is not causation. It's impossible to know whether a host of other things that the researchers didn't ask about might explain why life meaning and boredom are associated.

The question of causation is difficult to test. It is not ethical, or possible, to experimentally alter the meaningful events of a person's life, such as weddings, funerals, and the birth of children. However, it is possible to ask people to think about a time in their past when they felt either a high or low degree of life meaning. Doing so temporarily biases their thoughts and feelings about life meaning, and we can then look to see if this impacts in-the-moment feelings of boredom.

We brought people into the lab and gave them a detailed definition of life meaning. Next, we asked them to recall, and briefly write about, a time in their lives that was particularly meaningful to them. Others were asked to recall and write about a time that was meaningless. Then, after temporarily altering peoples' feelings of life meaning, we measured their levels of boredom. As expected, people who were asked to remember a meaningless time reported higher levels of boredom compared to those who were asked to remember a meaningful time. It seems a lowered sense of life meaning *can* cause boredom.[12]

When we have a sense of meaning and purpose in life, options for engagement with the world are evident and compelling. On the contrary, if we lack meaning and purpose, the value or significance of our options for action start to fade. It becomes difficult to find a reason to do one thing over another when there is no reason to do anything in particular. Meaning tells us to do things because those things are important. Without life meaning, we are directionless, caught in the desire conundrum, and bored.

Taking a different tack, we made people bored, then asked them about their sense of life meaning. It turns out that making people bored did not make them feel as though their lives were meaningless.[13] Sitting through a boredom-inducing four hours of a Christmas recital at your child's school in order to catch the five minutes that your son or daughter is on stage does not diminish your sense that parenthood is meaningful and worthwhile. Thus, diminished life meaning may cause in-the-moment feelings of boredom, but the same is not true in reverse. On first glance, this might seem inconsistent with our earlier claim that boredom is associated with the feeling of purposelessness. This contradiction disappears with the distinction between life meaning and situation meaning. We found no evidence that boredom changed people's sense of life meaning, despite the likelihood they felt that the situation they were in was meaningless.[14]

People who lack meaning and purpose in their lives report that they often feel bored. Lacking meaning and purpose in life can actually cause boredom. But being bored doesn't mean you will inevitably feel your life lacks meaning. This makes intuitive sense. It's hard to imagine why any given episode of boredom would result in changes to one's thoughts and feelings about life more generally. However, it might be the case that chronic experiences of boredom can, over time, shift people's views of the meaningfulness of their lives.

So, research findings both corroborate and refine earlier thinking about boredom and meaning. As we've argued throughout, we see boredom as a signal that our mind is unoccupied and that we are caught in a desire conundrum—in short, we are unengaged. Our own view is that a lack of engagement is more central to boredom than a lack of meaning.[15] Nonetheless, activities that do not occupy our mind and that do not flow from our desires are typically experienced as worthless and lacking in value; when we're bored we experience the situation as meaningless. That assessment may then push us to search for something that is meaningful.

Meaning Found

What do nostalgia, generosity, extreme political beliefs, and aggression toward people we judge to be different from us all have in common? They are things people turn to for relief when bored. At first glance, it is difficult to see why boredom is related to such a diverse range of outcomes. But the key to them all is meaning. It has been known for some time that people will use these outlets as a defense against the threat of meaninglessness.

Van Tilburg and Igou conducted a program of research to see if people will engage in these meaning-regulating behaviors if they are made to feel bored in the lab. If so, they reasoned, this would provide compelling evidence that boredom is not simply associated with feelings of meaninglessness but also induces a drive to *regain* lost meaning, which can express itself in a variety of ways.

For example, that drive to find meaning can push us into states of nostalgic reverie. People who were asked to recall any memory they wanted after doing a really boring task brought to mind more nostalgic memories compared to people who completed a less boring task. Meaning was the linchpin that brought boredom and nostalgic memories together. People who were bored were also more strongly searching for meaning, and this search was related to an increase in nostalgic memories. It would not make much sense when searching for meaning if you thought back to what you had for breakfast that day.

Instead, you tend to recollect those pivotal moments in life, such as when you met your life partner, for example. These memories have a greater sense of personal relevance and meaning to you. Finally, to complete the picture, when recalling nostalgic memories, we see an increase in feelings of meaningfulness.[16]

Being bored can also make us more generous. Research shows that people are more willing to make larger contributions to a charity when they have just completed a boring task compared to when they have completed an interesting, engaging task. Furthermore, these bored individuals were willing to give more if the charity was notably effective. In contrast, non-bored individuals' willingness to contribute was not impacted by the effectiveness of the charity. Perhaps bored people were particularly interested in *effective* philanthropy because their boredom drove them to seek out prosocial behaviors to reestablish their lost sense of meaning. At the very least, it shows that they were not simply seeking stimulation or trying to buy their way out of the boring ordeal. Rather, they were taking pains to make sure their money was actually making a difference in the world.[17] Maybe Milton Berle was on to something when in 1949 he launched the first-ever telethon—an ordeal so boring that it has raised millions of dollars for charity. Unfortunately, however, the pursuit of meaning does not always lead to positive ends; in fact, it can have a decidedly dark side.

Jingoism—extreme patriotism involving an aggressive stance toward outsiders—is arguably the cause of much suffering in the world today. And boredom may be one cause of this devastating social phenomenon. People who were made to be bored in the lab, compared to those who were not bored, expressed more positive attitudes toward the names and symbols of their own culture. They also assigned stiffer punishments to people they did not identify with (i.e., those from a different cultural background) and more lenient punishments to those from their own culture. It appears that boredom can drive people to regain a sense of meaning, which in turn causes a shift in behavior—a shift that can be directed toward strong symbols of identity.[18]

Boredom may also push our political views to the extremes. When self-identified left wing/liberal and right wing/conservative university students were asked to complete a boring task, they subsequently rated themselves as more extreme on the political spectrum compared to similar individuals who did not complete a boring task. Being bored did not, by itself, make people more right wing or left wing. Boredom simply accentuated the preexisting political differences between people.[19] A strong, uncompromising sense of identity certainly provides a sense of meaning and purpose in life and helps keep boredom at bay. We are not claiming that the ever more polarized (and polarizing) world of politics is primarily the result of an epidemic of boredom. But boredom and the attempt to find meaning in identification with a tribe might indeed be part of the story.

The push toward establishing some sense of meaning casts boredom—at least in terms of political extremism and the vilification of the "other"—as a driver of tribalism. Making matters worse, we know that people who often feel bored are also more likely to admire and be inspired by heroes as they search for something meaningful in the world.[20] This is a potentially dangerous cocktail. Worldwide, at any given moment there are many people who are bored. They may feel as though there is no point to what they are doing, and they may be urgently looking for answers. A charismatic leader, one whose rhetoric polarizes people into "us" and "them," may gain dedicated followers, despite his (such leaders are most often male) obvious foibles. From there it is a short step toward ever-greater extremes in political and world views. The leader has given their lives meaning, and they will not relinquish it easily. Unfortunately, both history and current events are replete with such frightening scenarios. The theologian Nels F. S. Ferre perceptively noted that "A man who experiences no genuine satisfaction in life does not want peace. Men court war to escape meaninglessness and boredom, to be relieved of fear and frustration."[21]

In fact, a recent sociological analysis implied that exclusively seeking peace is short-sighted. To minimize war, we must ensure that people are able to author their own lives and find meaning. Otherwise boredom will flourish and, in turn, give rise to a fascination with violence and the glorification of war. Clearly, boredom is not enough by itself to trigger a war, but it may set the stage and give permission for aggression that inflates a flagging sense of meaning.[22] When bored, we cast about looking for something that will make us feel as though our lives have purpose. To "fight for King and country," to blame immigrants for all manner of social ills, or to join well-defined groups that denigrate others, all fit that bill. Boredom isn't a judge and jury and cannot warn us that these attachments may be morally indefensible or have catastrophic outcomes.

Boredom ignites the desire to reestablish meaning. Whether our attempts to do so lead to positive actions or outcomes (nostalgic reverie, philanthropic largesse), or push us toward destructive pursuits (extreme political views, aggression toward others) is ultimately up to us. The onus is on us to respond to boredom in ways that are good for society and ourselves both in the short and long term.

Meaning in Progress

So far, we have emphasized meaning as a *product*; that is, we lose or find meaning in specific situations or in our lives more generally. This might be a personal meaning that we realize as an individual or a collective meaning, rooted in larger social structures, in which we participate in as an individual.[23] A deeper understanding of boredom is possible when we consider its relation to the process of *meaning making*. Although lacking meaning is a problem, merely having meaning may not be enough, either. Over time, what was once fresh and compelling can become stale. Personal convictions and life projects need to be reevaluated, changed, or reaffirmed. Like an intimate relationship that needs continual cultivation to thrive, meaning making is an ongoing process.

We are capable of making meaning even where none objectively exists. The most benign (and perhaps banal) example of meaning making is cloud watching, where we see faces and mythical animals in the shapes that drift and morph across the sky. Even more extreme, if you stare at white noise (such as the "snow" on a television) for long enough, most of us will eventually start to see shapes—shapes that simply aren't there.[24] This shows the human brain to be a meaning-making machine, seeking meaning in everything we experience. In a study from the 1960s, researchers asked people to read proverbs after undergoing a period of sensory deprivation. In one instance, the proverbs were written correctly, and in another, the words of the proverbs were mixed up. Boredom was highest when reading the unjumbled proverbs, the situation in which there was no challenge to meaning making. In contrast, when there was an opportunity to engage in the process of making meaning from the jumbled words, people were less bored.[25] When we stop collaborating with the world to create meaning, we are more likely to be bored. There are two sides to the coin of meaning as a process. On the one side, a situation must invite and support our participation; it can't be predetermined, fixed, or chaotic. The reverse side is about us; we must actively participate rather than passively receive.

Consider, for example, Figure 7.1. What do you see in the image?

Figure 7.1. The rabbit-duck illusion. Created in 1892, the image is ambiguous. While the physical image does not change, our interpretation of it can. If you can't see it yet, the duck's beak is also the rabbit's ears.

The image is one you have probably seen before. It highlights the fact that we bring something to the process of meaning making. Although the physical image does not change, we can interpret it as either a duck (beak facing to the left) or a rabbit (snout to the right, ears to the left). Contrast this with the symbols used to denote male and female public toilets. These are unambiguous and require little on our part to interpret them.

A situation is boring if it can be engaged with and understood in only one specific way. Sort of like the difference between a toy—a spinning top, let's say—that permits only one form of play versus modeling clay, with boundless opportunities. Any situation that leaves no room for us to shape meaning is quickly felt as meaningless and boring. In other words, when the meaning of a situation or object is fixed and predetermined before we arrive on the scene, what you see is what you get, with no possibility of anything more. And what you see is what everyone else sees. And what you see now is what you will always see tomorrow and the next day and the next.

It could be said that boring situations fail to abide by the Japanese garden design principle of *miegakure*—hide and reveal. The Japanese promenade garden is meant to draw you in with the promise of more. There is no single vantage point from which the whole garden is revealed. Instead, aspects of the garden are discovered or unveiled through the observers' movements through it. It is the promise of discovery that makes the garden enticing, like a burlesque show that is more engrossing than explicit nudity. To be bored is to lose the allure of the possible, to feel stuck in a dense, interminable present. The flow of time stops so that there is no momentum to carry into an unrealized future; instead, the future is just more of the present.[26]

Whether a situation offers a predetermined or fixed meaning is only half of the story. Imagine attending a university lecture on quantum mechanics. You know nothing about physics. A jumble of technical words washes over you to no effect; nothing makes sense. You simply can't understand what is being said. In short order you find yourself mind-numbingly bored—out of your skull with boredom. The flip side, then, to predetermined, fixed meaning is that extremely complex information also impedes meaning making. These two ways of preventing meaning as a process could be labeled *redundancy* and *noise.*[27]

Some redundancy is good. If every public toilet chose dramatically different symbols for signifying male and female, there would be some serious confusion. But if the movie we are watching replays the same scene over and over again without variation, most of us will wind up bored. Variety is said to be the spice of life because it attracts our attention and pushes us to expand our knowledge. But too much variety eventually becomes noise, so much "sound and fury, signifying nothing."[28] The lecture on quantum mechanics is fine for the expert, but for the novice it is just too far beyond anything we are familiar with. It is 0 percent redundancy and 100 percent variety. It's so novel we can't make sense of any of it and it is relegated to mere noise. Either too much redundancy or too much noise prevents us from effectively making meaning (an idea we unpack in Chapter 8). Situations that do not draw us in and allow for meaning making are boring. More than that, these situations render us impotent.

Boring situations impoverish us as meaning makers. Svendsen expressed it well: "Man is a world-forming being, a being that actively constitutes his own world, but when everything is always already fully coded, the active constituting of the world is made superfluous, and we lose friction in relation to the world."[29]

Superfluous is a key word in that quotation. Any situation that is fully coded renders our own capacity to interact with the world unnecessary. Boring situations objectify us, rob us of agency, and essentially make us interchangeable with any other observer. We can't customize the experience in any meaningful way. The icons for males and females on public toilets provide a good example: there is no need for us to make interpretations because we do not create the meaning.

The fact that we are incapable of participating in the meaning-making process makes us feel not only superfluous but lacking in control. We lose the sense of agency we first spoke of in Chapter 2, and we become passive recipients of meaning rather than active creators. The spinning top does only what it was designed to do, and we must accept that limitation. Worse than that, we are prevented from imbuing the toy with any other meaning that might match our needs, desires, or intentions. As Andreas Elpidorou, a philosopher from University of Louisville, puts it, "the world of boredom is, in a sense, not our world, it is not the world that is in line with our projects and desires."[30]

As humans, we have an inherent ability and need to make meaning, and boring situations deny us the chance to exercise this meaning-making capacity—it is in this sense that boring situations impoverish us. Indeed, in a very real sense we become unnecessary. This experience may, in part, explain the indignation and contempt we feel when we are bored. Boredom is an affront to our personhood.

Meaning Uncertain

Boredom alerts us to the absence of meaning. And when things work well, boredom's signal guides us toward activities that utilize our abilities, express our passions, and ultimately help us find meaning in life projects. But it does not always work out so well. Boredom is an inefficient spur to meaning. In fact, the absence of boredom does not ensure the presence of meaning.

There are ways of dampening boredom that do not involve the creation of meaning, activities that are ever present and beguiling. It may be that technology has short-circuited the adaptive relation between boredom and meaning. Binge-watching Netflix or wasting hours on Candy Crush are not likely optimal tools for creating meaning. Contemporary society is full of easily accessible, quick acting, and temporary relief from boredom. But it is reasonable to wonder if our attachment to these forms of relief is actually making boredom an even more pressing concern. Are we, as has often been claimed, in the midst of a boredom epidemic?

CHAPTER 8

AN EPIDEMIC IN THE MAKING

• • •

The alarm sounds at a time you never thought decent for a Saturday morning. You haul yourself out of bed and head downstairs to get your 7-year-old son moving. He's sitting on the couch, a tablet illuminating his face as he plays Minecraft.

You pause for a moment wondering whether they still show *He Man, Masters of the Universe* cartoons on TV anymore. Or better yet, reruns of *Rocky and Bullwinkle*. Your son probably wouldn't like them anyway. He sure likes that damn tablet!

"Time for breakfast, buddy, let's go. We've got soccer to get to."

He doesn't look up but sort of grunts an acknowledgment to let you know he heard.

Later, while he and a group of 6- and 7-year-olds swarm around a moving soccer ball, you pause, looking around at the parents gathered to watch their children's exploits. They all have a smartphone in hand. Periodically they look up to yell encouraging words at their child before plunging back into whatever online activity compellingly holds their attention. What is it: Candy Crush? Snapchat or Instagram? Plain old texting? For the truly old school, perhaps it's Facebook? There's no shortage of things to drag their attention away from the little ones running up and down the field.

> Never before have we had entertainment so readily available, yet some say boredom is at an all-time high. You wonder, with such easy access to novelty videos of cats doing dumb stuff, how could anyone complain of being bored? Unless, of course, we're all actually suffering from information overload? Or worse, our engagement with the Internet is a false dawn, a retreat from true engagement.

• • •

"It is not at all unusual to have the television or radio on while reading a newspaper, eating breakfast, and now and then talking to somebody. We may also carry a watch or pocket pager whose beeping signals remind us to move along to new stimuli. We may also carry a cassette recorder ('ghetto blaster') as we walk down the street—adding noise for everyone else—or wear a 'Walkman,' giving continual entertainment in our own sonic cocoon."[1] Orin Klapp wrote about our busy, distracted lives—our "sonic cocoon"—in 1986! Replace the anachronistic "ghetto blaster," "Walkman," and "pocket pager" with a smartphone and the claim seems at home in our present day.

Then as now, we are overloading ourselves with information, and at some point, too much information becomes noise. When that happens, we disengage and boredom ensues. Or so the story goes according to Klapp, a sociologist working at Western University at the time he wrote this. Just imagine, if Klapp was concerned about the effects of information overload in the mid-eighties—the era of big hair, tight jeans, synth pop and glam metal—what might he make of this century's twenty-four-hour news cycles, vitriolic debates executed 140 characters at a time with thousands of followers, and the world at our

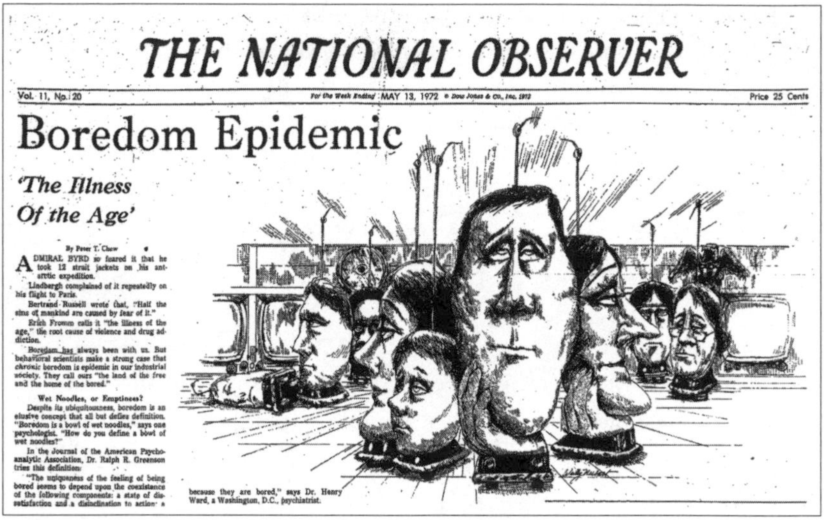

THE NATIONAL OBSERVER

Vol. 11, No. 20 For the Week Ending MAY 13, 1972 © Dow Jones & Co., Inc. 1972 Price 25 Cents

Boredom Epidemic

'The Illness Of the Age'

By Peter T. Chew

ADMIRAL BYRD so feared it that he took 12 strait jackets on his antarctic expedition.

Lindbergh complained of it repeatedly on his flight to Paris.

Bertrand Russell wrote that, "Half the sins of mankind are caused by fear of it."

Erich Fromm calls it "the illness of the age," the root cause of violence and drug addiction.

Boredom has always been with us. But behavioral scientists make a strong case that *chronic* boredom is epidemic in our industrial society. They call ours "the land of the free and the home of the bored."

Wet Noodles, or Emptiness?

Despite its ubiquitousness, boredom is an elusive concept that all but defies definition. "Boredom is a bowl of wet noodles," says one psychologist. "How do you define a bowl of wet noodles?"

In the Journal of the American Psychoanalytic Association, Dr. Ralph R. Greenson tries this definition:

"The uniqueness of the feeling of being bored seems to depend upon the coexistence of the following components: a state of dissatisfaction and a disinclination to action; a

because they are bored," says Dr. Henry Ward, a Washington, D.C., psychiatrist.

Figure 8.1. This headline, from the now-defunct *National Observer,* was published in 1972. The article claimed that we were experiencing a boredom epidemic because supercities, superhighways, supermarkets, and supercorporations all represented an unchanging monotony. The cartoon has humans not on a treadmill but in an endless bumper-car ride. The expressions—heads resting on hands, vacant stares—is reminiscent of Toohey's (2011) treatise on the depiction of boredom in art.

fingertips on devices barely larger than a credit card! Klapp wasn't the only one of his time and earlier to suspect that we were on the cusp—or even in the midst—of a boredom epidemic (Figure 8.1).

This concern that information overload may be problematic for humans and, more specifically, that advances in technology are the cause of these problems, is an age-old debate. In Plato's *Phaedrus,* Socrates lamented the advent of writing, suggesting that writing "will create forgetfulness in the learners' souls, because they will not use their memories; they will trust to the external written characters and not remember of themselves." Ironically, we would know nothing of Socrates' philosophy had Plato not decided to write it down. As noted earlier, William James, the nineteenth-century philosopher and psychology's fount of quotations, lamented that "An irremediable flatness is coming over the world"—a flatness he associated with increased *amount,* but not *quality* of information. Shortly after his lament, in the early part of the twentieth century, information overload was raised again as a concern by Siegfried Kracauer, a German writer, sociologist, and cultural critic, in relation to the explosion of print media. His claim was that more news did not represent more information and simply confounded our ability to sort signal from noise—essentially Klapp's argument years later.[2]

There is little to no data to suggest that we are in the middle of an explosion of boredom. The kind of data we would need—longitudinal data from the same individuals across many decades—simply aren't available. But if information overload is a condition ripe for boredom, then the possibility arises that with so much information so easily available, boredom may be more problematic than in previous times. Perhaps it is not so much an epidemic as a distinct manifestation of an age-old problem. Boredom signals the need for engagement, but will any kind of engagement do? Are we really sated by the hours we spend on Instagram or playing Candy Crush? We think not, but before we tackle the issue in more depth, we outline Klapp's argument in a little more detail. If we forgive the anachronistic details of ghetto blasters and Walkmans, there is much to learn from his account, now more than three decades old, of boredom in the modern age.

Sorting Signal from Noise

Klapp would suggest that it is the lack of information that makes monotony boring. Monotonous circumstances by their definition contain little to no new information, and humans, according to this account, have an inherent need to seek new information. This is not sensation seeking in the sense of the adrenalin junkie's need for thrills but a curiosity driven by the need for discovery. But what if there is too much variety? Here, according to Klapp, we are unable to extract a meaningful signal from the noise. As the ticker tape of news items scrolls across the bottom of the screen, a talking head tells us about a specific story. Insets to the right show changing traffic conditions on the major highways of our city plus weather forecasts at hourly intervals. How are we to make sense of the sheer onslaught of information? Boredom may well be one outcome of the imposing wall of information we encounter every day. As Klapp puts it, boredom "arises when [the] pace gets faster, change lacks meaning, and movement lacks arrival."[3] Things are constantly happening, but we struggle to make sense of them. And it is certainly true that the ease of access to information in our age at least gives the impression that things are changing at an ever-increasing rate.

So at both ends of this information spectrum—too little or too much—we have a crisis of meaning (Chapter 7). With too little information to satisfy and with little to no change from one moment to the next, things become unbearably monotonous. With too much information to wade through we may feel like we're constantly in motion—from one tweet to the next, from one funny cat video to the next, from one so-called "breaking" news story to the next—without ever pausing to figure out what, if anything, all of this information means to us.

This account of boredom as a function of information overload in a modern age casts the experience in terms of our need to discover and make meaning, and in so doing, express our agency. Not all repetition is bad. Familiarity anchors our understanding. We find a piece of music intriguing when it plays with our expectations—it is at the same time both familiar and surprising. Similarly, not all variety is good. A Schoenberg symphony is chaotic and impenetrable to someone unfamiliar with atonal music. Extreme redundancy and extreme variety both prevent us from getting an intellectual foothold so that we can perceive something new and meaningful. Optimal engagement then depends on finding a balance between repetition and variety—yet another kind of Goldilocks zone. In this context, boredom functions as a signal that mediates the oscillation between redundancy and variety. This fits with our claim that the boredom signal is not itself problematic. It is how we respond to it that leads to positive or negative outcomes. It also suggests that the highly boredom-prone individual lives at the extremes of information processing—either mired in monotony or overwhelmed by novelty—in either case unable to find the "just right" zone.[4]

Klapp outlines nine potential routes to information overload and subsequent boredom,[5] but ultimately it is the sheer amount of available information that is most relevant to us here. Information comes at us at an ever-increasing rate. Simply being awake and engaged with the world can feel like drinking from a fire hose. It's too much—too much mental stimulation, too little meaning, and after the rush of stimulation, ultimately boring. It takes time and effort to make sense of information, to find meaning among the bits and bytes. In other words, meaning making is slow, whereas information accumulation can be rapid, leading to what Klapp calls the "chronic indigestion of information."[6] Meaning *making* requires synthesis of the different sources of information and interpretation of what it all means. Yet *acquiring* that information can be a simple matter of passive intake. When so much information bombards us in the Internet age, we may respond to these challenges of synthesis and interpretation by adopting shallow levels of processing. We may rely on tweets to avoid the harder work of making deeper sense of the deluge. Or we may simply shut off altogether. Either way, we become bored.

Ultimately, the explosion of information in our digital age—an explosion of a magnitude that few of us could have anticipated—compounds, rather than solves, the challenge of meaning making. If boredom arises in circumstances that thwart our ability to engage and make meaning, then we might expect that our own age will be awash with bored people. In other words, has our current Age of Information—or indeed, an Age of Information Explosion—made the task of meaning making, and by extension the avoidance of boredom, far more difficult?

It is worth stating once again that there is little data to speak directly to this question. To measure changes in expressions of boredom over time we would want to ask a large group of people the same questions over a long period—a longitudinal study to see how individuals' perceptions change with time and to see how each new generation compares to the previous one.

To know whether the Internet, smartphones, social media, or the Twitterverse are the key culprits in any hypothetical rise in boredom is even more challenging. Find a group of people who do not use such things to function as your comparison group—a group you would hope shows no rise in boredom, given their freedom from the information onslaught? The Amish might provide such information, perhaps, but such a group is so culturally different from mainstream society that the comparison is really not appropriate. So, unfortunately, we don't yet have the necessary data sets to know if boredom is on the rise and the Internet is to blame.

What we can do is turn to other sources of data to help us make inferences about what might have changed over time. Things like Gallup Polls can be helpful, although the questions asked change with the exigencies of those seeking the information. Assessing frequencies of use of terms like boredom in literature or other sources can also give hints at changes over time. Indeed, an analysis of this kind from Klapp showed that the words boredom, routine, and monotony were used 2.5 times more often in 1961 than they were thirty years earlier in 1931.[7] Boredom alone was used ten times more often over that same time span. In Gallup Poll data taken around the same time (1969), around 50 percent of respondents claimed their lives were routine or even pretty dull. Alas, this question was asked only in that poll, so we can't determine any upward or downward trend.

Regrettably, we simply don't have enough to go on to definitively say whether or not boredom is on the rise. But if we accept the notion that increased information and ease of access to that information is on the rise, then we can start to examine other consequences of this increase in information that relate directly to boredom.

It seems incontrovertible to us that information, and our access to it, has dramatically expanded over the past century and even over the past two decades. Buckminster Fuller—the architect known most for creating geodesic domes—coined the concept of a knowledge doubling curve (although we might want to distinguish between knowledge and information).[8] According to this notion, up until 1900 human knowledge had doubled roughly every century. By the middle of the twentieth century that rate had gone up, doubling every twenty-five years, and now may be doubling as rapidly as every twelve months. Specific forms of knowledge accumulation likely accelerate at different rates. Scientific information is thought to be doubling only every eight to nine years, although the rate of change throughout the twentieth century and into the twenty-first is exponential.[9] And the ease with which we access that information has also sped up unbelievably. Some of us remember photocopying articles from journals in the library or even reading material on microfiche.[10] At the time that Klapp was writing his treatise on boredom, information could be transmitted at a rate of 256,000 bits per second. Now that rate of transmission is closer to 100 megabytes per second.

In Klapp's terms, boredom should inevitably be on the rise, given this explosion of information. We would, by necessity, either retreat into less noisy environs or try desperately to be heard above the noise, what he referred to as "ego-screaming," a response to boredom provoked by information overload. It's hard to avoid simply accepting Klapp's theory as fact, given the popularity of social media. The sharing of mundane events with followers in a desperate search for "likes" seems a clear cry to be heard above the noise. Add to this the Twitter wars, ubiquitous cat videos on YouTube, and the rise of online trolling and it seems obvious that we are in part responding to the explosion of information by trying to shout louder than everyone else. But it is important to point out that this remains a logic argument, one that is wanting for empirical support.

It is not simply information per se that would be responsible for any rise in boredom but the medium itself. As we've already hinted at, perhaps it's the Internet, social media, and smartphones that are to blame. Here, at least, we can turn to some recent evidence to interrogate these claims.[11]

The Allure of the Internet

The origins of the World Wide Web are complex, involving many research groups across the United States and Europe, beginning in the late 1960s. The first public Internet service providers began operating in the late 1980s, and within a decade the availability and usage of the Internet as we know it had exploded.[12] In lockstep with this explosion, by the late 1990s a controversial proposal for the diagnosis of a new mental illness arose—that of Internet addiction.[13] The addiction was defined in terms normally reserved for substance abuse disorders (addictions to alcohol or drugs, for example). People who spent more than thirty-eight hours a week on the Internet, who felt anxious when not connected, and who experienced problems in their daily lives as a result of Internet use were considered to be addicted. Early research showed that Internet use and boredom were indeed linked, and it was suggested that turning to the Internet for engagement was a *response* to being bored.[14] For example, in young people aged 16 to 19 the prevalence of Internet addiction appears to be driven by leisure boredom (Chapter 4). Finding no other outlets for satisfying engagement, these youth turn to the rabbit hole of the Internet, and between 4 and 12 percent of them, depending on the study, struggle to find their way back out. When bored we seek outlets to remedy or eliminate the feeling. The Internet dramatically expanded the available *stuff* to engage with, not to mention the ease with which these can be accessed.

In the early days of widespread public access to the web, Internet use required a computer. This has clearly been supplanted by the ubiquitous smartphone. Researchers are beginning to define what they consider to be problematic smartphone use. Beyond simple increases in frequency of use, smartphone addiction symptoms include feelings of anxiety and hostility when unable to access it—that is, feelings of withdrawal. Using a statistical technique known as Structural Equation Modeling, which examines the relationships between different variables, Jon Elhai and colleagues from the University of Toledo show that boredom predicts problematic smartphone use.[15] That is, the more bored you are, the more likely your attachment to your phone will be unhealthy.

The argument here is a tricky one. It is not as simple as Klapp's claim that increased information leads to overload and ultimately boredom. Rather, the Internet and smartphones offer beguiling relief to the problem of boredom but ultimately let us down—and likely make things worse in the long run. The Internet does, like any addictive activity, provide quick, effective, and easy relief in the moment. Boredom is kept temporarily at bay while watching YouTube, but it is a crutch that is ultimately unhelpful and unhealthy.[16] A vicious cycle ensues. If we are unable to occupy our minds and become engaged, the Internet becomes a place of refuge from boredom. That refuge is ultimately meaningless, characterized by vast quantities of information that afford only shallow engagement. Disconnecting from the Internet leads us right back to the challenge we first faced. What to do? Social media, smartphones, and the Internet are *engineered* to capture and hold our attention and provide short-term relief from the doldrums; it's no wonder we are held captive by a vicious cycle we can't overcome.

The intended goal of surfing the web is often given as "passing time."[17] Not unlike the description given by prison inmates of "killing time" or "doing time," this represents the feeling that there is nothing better or more meaningful to do. Killing time in this way highlights something important about the Internet as a false dawn in the quest to avoid boredom: we want something to do, but it can't just be anything. Boredom, as we define it, is in part driven by the desire to do something that allows us to optimally deploy our skills and talents—something that when completed, we can look back on with a sense of satisfaction. The Internet, and perhaps social media more specifically, at least give us something to do, something to occupy our time. But it is at best a simulacrum[18] of agentic engagement; never fulfilling our deeper need to be self-determined and doing something of value.

So, too much information is ultimately boring, and the Internet and smartphones offer a simulacrum of the true engagement and connection we crave. To borrow a description from Leslie Thiele, a Professor of Political Science at the University of Florida, we have offered a speculative account of how boredom and technology "collude."[19] Each works with the other for its own ends. Boredom pushes us into the arms of technology so much so that we cherish it and endlessly seek to develop new and better forms of distraction. On the other hand, technology leaves us ultimately unfulfilled, ensuring boredom's continuance.

We can speculate about other "back room deals" technology has made to keep us bored and looking for the relief it purports to provide, beyond offering a beguiling, yet ultimately hollow engagement and connection. For Thiele a key problem with technology is that it has shifted the way we think about the world and, in so doing, made us ripe for boredom. In this account, our technological world view primarily values *efficient function*—what is the utility of this thing and how well can it perform? For Thiele, this changes the relationship with time for technology's end users—us. Rather than simply "dwelling in time,"[20] we try to master and manage time, all at the altar of improving efficiency and productivity. Time becomes a problem to solve, and once we make that shift, unfilled time becomes inefficient, unproductive, and ultimately boring. Filling any unfilled time creates another vicious cycle. Time must be filled, but not by just anything—technology has led to what Thiele calls the "routinization of novelty." We become driven by the imperative to constantly innovate, craving novelty but perpetually raising the bar on what constitutes something new and interesting. What was engaging yesterday fails to satisfy today, ultimately leading to a hamster wheel of never-ending searches for novelty. Essentially, we become numb and bored with what's on offer.

The dominant technological worldview of today may also change how we think of ourselves in ways that make us more prone to boredom. Perhaps the most devastating shift is from creators of meaning to passive consumers of experience—as containers to be filled rather than agentic sources of meaning (Chapter 7). If we can be the kind of thing that can be filled with ever varied and compelling experience (or at least a constant stream of new memes), then we also have the potential to be empty.[21] We are propelled toward ever more consumption just to stay one step ahead of boredom. The less we see ourselves as agentic meaning makers, the more our capacity to be agents atrophies. Georg Simmel, the German sociologist, remarked at the beginning of the twentieth century that being surrounded by technology is like being "in a stream . . . [in which] one needs hardly to swim for oneself."[22] Constantly carried along by a fast-moving stream, we forget how to swim for ourselves. We stop asking what really matters, and we lose touch with our inner desires. Put another way, technology is unrivaled in its ability to capture and hold our attention, and it seems plausible that our capacity to willfully control our attention just might wither in response to underuse.

Connectionless Connections

While we are indeed "connected" when we're online, it is a less satisfying form of connection. In Elhai's study it was the way people used their devices that mattered and how they felt when they were unable to access it. Ultimately, problems arise when our connection to technology stands in for real social interactions.

Jean Twenge, a social psychologist from San Diego State University, shows that today's teens—who she labels iGen to highlight the fact that they are the first generation to be born into a fully online world—spend less time with their friends and more time on their phones than generations before them.[23] And there are consequences. Contrasted with the teens who are spending more time with friends and playing sports, those who spend more time on the Internet, playing games or engaging deeply with social media, also report lower levels of happiness. They're online, but they're disconnected. Some recent studies of social media use have shown that even abandoning Facebook for as little as two weeks can improve happiness.[24]

Another indication that the Internet is a form of illusory connection—or connectionless connection—comes from closely observing behavior while people are online. One study that used laptop cameras to observe behavior suggested that on average people changed what they were doing every nineteen seconds![25] You might be constantly doing something, but it is hard to imagine being fully engaged if you change what you're doing three times a minute. Pop-up notifications and the never-ending possibility of things to pay attention to leave us distracted and mentally exhausted so that we can't deeply connect with any one thing for an extended period of time. When distracted and unable to engage with the task at hand, we will soon find ourselves overcome by boredom.[26]

Perhaps the ultimate in connectionless connection comes from Internet porn, which arguably takes up a lion's share of the Internet's bandwidth. Those who turn to Internet porn regularly may be suffering from what has been labeled sexual boredom.[27] More prevalent in males than females, sexual boredom is related to the need for sensation seeking—this time the thrill-based kind—or just anything different from one's regular encounters. Clearly associated with discord within relationships, sexual boredom is also related to trait boredom proneness and a higher prevalence of solitary sexual acts and viewing of Internet porn. The same cycle of unhealthy reinforcement may be at play. Unable to satisfy your sexual desires in real life, you turn to the Internet. This temporarily sates the need but ultimately falls short of the true connection you were seeking. You turn back to the real world, but any failure to satisfy sends you back to the ease of access and breadth of stimuli that is Internet porn. This cycle again highlights our point that the boredom signal itself is not the enemy—it is our response to it that leads to different kinds of engagement and their attendant problems. Connection without meaning, sensation without depth, information without context, all are unlikely to ever fully alleviate boredom.

A common thread in our Internet-fueled obsessions is isolation. In Chapter 6 we spoke of environmental isolation, whether by choice to embark on expeditions to extreme locales (outer space, the Antarctic, and so on), or enforced as part of a punitive legal system. Here, the type of isolation is more social. Failing to satisfy our desire to interact with others may push some to the false panacea that is the Internet and social media. To be sure, connecting online undoubtedly has huge benefits. It allows families separated by distance to communicate, enables business partnerships to flourish across borders, and disseminates information more rapidly than oppressive governments can stifle it. All this and more is clearly beneficial to society at large and to most individuals. But for some, Internet

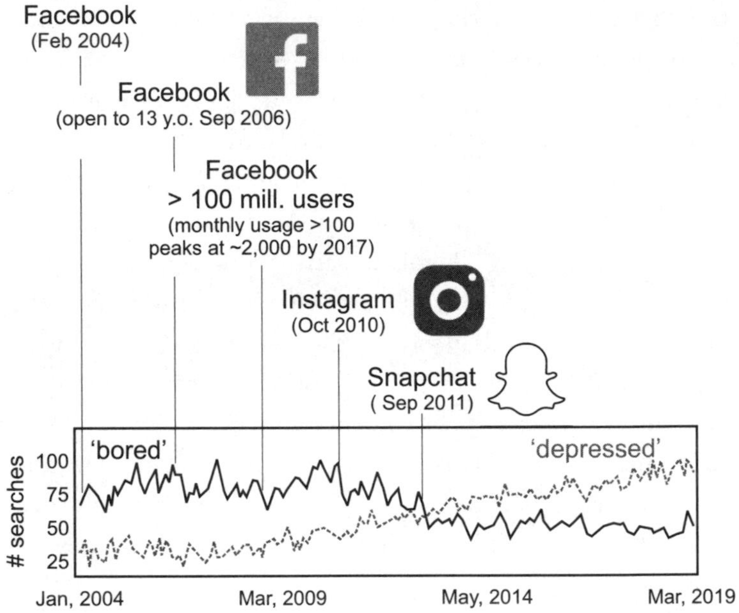

Figure 8.2. This graph charts the frequency of two search terms in Google from 2004 onward. The entrance and influence of particular social media platforms is noted.

connections may represent the only connections they have. This is where the Internet and perhaps more pointedly social media can become problematic. In all likelihood, it represents a new expression of an old problem. What we are claiming is that for some, the illusory connection of the Internet is an unhealthy response to the disconnection from the world that boredom signals.

To get just a glimpse into the illusory connection hypothesis we are proposing, we searched Google's database for the frequency of search terms used since 2004, when Facebook was launched. Searches including the words *bored* or *boredom* look like they're actually on the decline. But searches for the word *depressed* are on the rise (Figure 8.2). We may feel like we have enough to occupy our time, but what we occupy our time with may not be enough to deeply satisfy.

Lonely Hearts Club

Connectionless connections of the kind we suggest the Internet and social media are particularly adept at fostering may ultimately and ironically lead to more social isolation for some. We may feel connected more than ever, but time spent on Snapchat is time taken away from genuine personal interactions. Early in 2018, Theresa May, at that time the British prime minister, announced that her government was establishing a Minister for Loneliness.[28] A special commission to study loneliness suggested that some 9 million Britons—close to 14 percent of the population—felt lonely often or most of the time.[29] The newly minted minister was to be responsible for establishing policies to deal with the public health crisis represented by rising levels of loneliness. In the United States, studies suggest that loneliness is indeed a common ally of boredom.[30]

Although much of the media attention related to May's announcement (and the ensuing jocularity from late-night television hosts) centered on the elderly, a population in which loneliness and boredom go hand in hand, the concerns were broader than that. As Jo Cox, the British MP whose posthumous foundation publicized the initial study of loneliness in the United Kingdom, once said, "Loneliness does not discriminate." Neither does boredom. Social isolation in teens struggling to find their place, in new parents burdened by lack of sleep and adjusting to a new reality, in the mentally ill isolated by distress and impairment, in the elderly and their care givers encumbered by higher needs than most, loneliness and the attendant boredom, anxiety, and depression can strike anyone, anytime.

There is a key way in which loneliness and boredom are linked; both represent a form of disengagement from the world. We have argued that the state of boredom is a signal telling the individual that they need to become mentally occupied, to engage with some activity that allows them to deploy their skills and talents. A common outlet for the need to be mentally engaged is social interaction with our fellow human beings. Indeed, one recent study using experience sampling hinted that boredom was lowest when people were engaged in face-to-face activities.[31] For some, the response to boredom and isolation is to go down the rabbit hole of social media, the illusory connection easing the discomfort of boredom in the short term but ultimately failing to establish meaningful relations with others. Even worse, that failure to deeply satisfy our need to be socially engaged may even perpetuate boredom in the long run.

We began this chapter with the notion that boredom can arise from too much variety because it becomes impossible to extract a signal from the surrounding noise. And we noted that we are surrounded by an ever-increasing onslaught of information. We proposed other ways in which the Internet, smartphones, social media, and technology more generally might give rise to boredom and loneliness. We've outlined ways in which boredom and technology collude to keep us trapped in vicious cycles. It's important to note that these ideas remain speculative. The critical research studies have simply not been done to confirm these claims. But as we turn to these easy outlets to mollify the discomfort that comes with being bored we may inadvertently make it harder to find the connections we really need. It would be a stretch to claim that it is good to be bored. But we hope, over the course of this book, to have convinced you that it is good to at least *retain the possibility of boredom* and to learn how to respond to it in productive ways.

Boredom is neither good nor bad. It doesn't do the hard work of deciding for us *what* we should do to alleviate the experience. But without the capacity to recognize that we are bored, we run the risk of persisting in a maladaptive state of mental disengagement or deluding ourselves that we are fine as we become increasingly more tethered to the Internet. A key challenge for us, in the face of ever-present distraction, is to resist the allure of quick and easy mental engagement—junk food for the mind—that will ultimately fail to satisfy and will cycle us back to being bored all over again. Instead, we need to seek true antidotes to boredom that will satisfy our need to be mentally engaged.

CHAPTER 9

JUST GO WITH THE FLOW

• • •

When completed they would be the tallest buildings in the world. In a city forever trying to touch the sky, the Twin Towers of the World Trade Center would top them all. And ever since he had read about them in a magazine in France, Philippe Petit had had a dream. He would rig a wire between the towers and walk on it for the entertainment of those below—his *coup*.

He spent six years in the planning, noting security guard movements in the unfinished towers, obtaining aerial photographs, scouting for anchor points, making clandestine excursions to the top to run through every detail. He even went to the extreme of constructing a miniature version of the towers to work through any potential problems. And he practiced for hours upon hours, walking a rope only meters above the ground in preparation for doing it all 400 meters above the chaos of New York's streets.

The walk would last forty-five minutes. It was not simply a one-way walk across the wire. Petit intended to traverse the sixty-one meters eight times. He would hop from one foot to the other while holding his twenty-six-meter balancing pole, lie down on the wire, and wave to the gathering crowd.

Did Petit have a death wish? Was he a thrill seeker? On the contrary, on many occasions he has asserted that fear has no place on the wire, and while he acknowledged that the wire flaunts death, it was not death but life he sought. So if he had no death wish and the walk did not thrill him in a conventional sense, why would he do something so audacious? Perhaps for art's sake, but perhaps also to experience the extreme sensation of flow—extraordinary concentration, an imperviousness to distraction (Petit once claimed that on the wire you could hit him in the head with a plank of wood and he would not move!), and a sense that his actions were one with the world.

• • •

Alex Honnold is a preeminent mountain climber, specializing in what is known as "free-soloing"—climbing mountains like El Capitan in Yosemite National Park—3,000 feet of sheer granite, without the assistance of any ropes.[1] Most of us would see Honnold's climb as crazy, like Petit's walk on the wire, or maybe even providing evidence of a death wish. But that's not how either one of them describes the experience. Planning is paramount, fear has no place in the endeavor, and far from a death wish, the activity makes them feel more alive than ever. In some sense, for Honnold or Petit to succeed in what they do, they must reach a level of hyperintense engagement.

Given our description of boredom as restlessness born of the failure to engage, then maybe the state Honnold strives for represents a kind of opposite to boredom. But is hyperengagement, deeply intense concentration, the *only* opposite of boredom, or are there others? By examining states and feelings that represent, in some form, the opposite of boredom, we can deepen our understanding and ultimately be better equipped to respond effectively to the signal when it does arise.

There are a plethora of possibilities for what constitutes boredom's opposite. It might be excitement: the thrill of a rollercoaster ride, the anticipation of your team's impending playoff victory, that moment before you go on stage to perform. Turn the dial down a little and perhaps interest is boredom's opposite: engrossed in a novel, immersed in a complicated movie plot, even absorbed with a 1,000-piece puzzle. And what of pleasure? A four-course meal, an exhilarating concert, a night of passionate lovemaking? If not pleasure, relaxation? Resting poolside, walking through the park on a spring afternoon, "vegging out" in front of the TV at the end of a hard day's work? Might any or all of these be the opposite of boredom? Quite simply, yes. What is common across excitement, pleasure, interest, curiosity, and even relaxation is that when we experience such things, our mind is occupied and we *want* to be doing what we are doing—we are satisfactorily engaged with the world. Below we explore some of these potential opposites of boredom through the lens of engagement, starting with perhaps the most prominent candidate, one that Honnold and Petit likely feel when we presume they should be in the throes of great terror—the experience of flow.

Flow

Almost fifty years ago, Mihaly Csikszentmihalyi began developing a novel theory and methodology to address the "how" of human happiness.[2] Like boredom, what makes us happy is ultimately idiosyncratic. Collectors can pore over the objects of their passion for hours on end, content to explore minutiae others can't appreciate or even bear to think about. The *con-*

tent of what we do is less important than how we connect with it. This was Csikszentmihalyi's insight. The methodology he introduced is known as "experience sampling." Using diaries (that nowadays involve smartphone prompts), direct interviews, or more recently interruptions of ongoing tasks, Csikszentmihalyi simply asked people about their subjective experiences of the everyday.[3] From rock climbers like Honnold, to factory workers, to surgeons, to performing artists, the message was clear—people were most fulfilled when so mentally engaged that the rest of the world seemed to fall away. Csikszentmihalyi's participants often referred to this phenomenon as "flow."[4]

Although they do not always parse them in exactly the same way, researchers have consistently identified a number of necessary characteristics for flow:

- our skills and abilities must be up to the challenge;
- we need a heightened sense of control;
- we need well-defined goals and clear feedback on progress;
- our attention must be intensely focused;
- our awareness must be so tightly linked to what we are doing that we lose sight of ourselves;
- whatever we are doing must feel effortless;
- what we do, we do for its own sake—we are intrinsically motivated, and
- our sense of time becomes distorted.

Some of these characteristics are arguably preconditions for flow while others can be thought of as the outcomes of achieving flow.[5] Regardless, they are all important aspects of the flow experience and when present, they may protect us from boredom.

A key precondition for flow is a balance between what the moment demands of us and our ability to skillfully meet those demands (much like the other "Goldilocks" requirements we've already discussed). According to Csikszentmihalyi, flow occurs when that sweet spot is achieved—moments when we are in harmony with the situation. If we miss the mark on one side, discovering that we are not capable of meeting the demands of the task, we feel anxious. If we miss the mark on the other side, when a task is too simple for our skill set, we feel bored. Boredom pushes us to seek out bigger challenges, whereas anxiety warns us of the need to enhance our skill set. Together, these negative feelings point us toward flow.

A climb that each of us could make at the local gym, such as a 5.4, a low rating for a climb, would be facile for someone with the skill of Alex Honnold. It wouldn't even be effective as a warm-up. He would likely skip straight to a 5.9 or 5.10 (challenging for a novice but a mere warm-up for him). If we think instead about the novice climber, a 5.9 climb may seem more of an insurmountable challenge than a warm-up. In either situation there is a clear mismatch between the individual's skills and the demands of the task. We seek a Goldilocks zone where the match between the challenge level and our own skill set is "just right" to push our limits and lead us away from boredom and toward flow.

Recall our experimental demonstration of this need to fit challenge and skill to avoid boredom from Chapter 2: if we artificially let people win at the game rock, paper, scissors all the time, the task lacked any challenge. Far from getting in the zone, people found the experience to be mind-numbingly dull. As we've explained, however, this wasn't the end of the story. People who lost all the time against the computer opponent first felt frustration before becoming bored.[6]

In another study, also first discussed in Chapter 2, we had two groups watch different twenty-minute videos. In one, the ridiculous mime plodded slowly and repetitively through a rudimentary English vocabulary lesson, and in the other, the mathematician taught advanced computer graphics via impossibly complex math and charts. No matter which video people watched, they wanted to poke their eyes out by the end. They were equally bored when under- or overchallenged.[7] Counter to Csikszentmihalyi's model of flow, boredom occurs not only when we are underchallenged. Instead, boredom arises at both ends of a challenge / skill mismatch—when things are either too easy or too hard. This fact has been realized for some time in the education world. Reinhard Pekrun, research chair at Ludwig-Maximilian University of Munich, has shown that when school tasks far exceed students' ability, they disengage from the task and experience it as boring because it can't hold their attention.[8]

A second condition for flow is the sense that we are in control. If we think again of Alex Honnold free soloing a cliff face, we might imagine that to be terrifying. But that expected sense of terror arises from our own self-awareness that in such a circumstance we would be well and truly out of our element. The opposite is true for Honnold. To be sure, when he dangles from his fingertips hundreds of meters above the ground there is a chance he could fall to his death. But he doesn't feel that sense of impending danger while in the midst of the climb. Instead, he has the sense that every move has been well planned and that he is in control of his own actions. To him that's what makes climbing so absorbing and compelling. In a state of flow, there is a sense that any eventuality can be handled and that at each moment *we* determine what will happen next. In other words, we feel a profound sense of agency, a key factor that makes it a peak experience.

In contrast, boredom thwarts our agency. When we are bored, we have a diminished sense of control. The world happens to us, and we can't change it. As we've seen in Chapter 2, boredom cuts us down before we even get started, making us unable to say what we want to do, yet imposing upon us the need to do something. If flow is characterized by peak agency, then boredom is surely the valley.

In a follow-up to our rock-paper-scissors study we deceived people in a different way. This time we had them play against a computer opponent that played each option equally often—meaning you could only ever win 33 percent of the time. We told one group the truth—your opponent is playing uniformly and randomly so there's no way you can win more than one-third of the time. For this group there was no way they could exert control in the game, making continued play monotonous and boring. The second group we lied to, telling them their opponent employed an exploitable strategy, and if they could figure it out, they would win more often. This group was not bored at all, despite never winning more than a paltry 33 percent of the time. The mere prospect of gaining control—even when never realized—was enough to stave off boredom.

We've already mentioned that for Honnold to make a successful free solo climb he must plan it out in meticulous detail, just as Petit spent years planning his walk between the Twin Towers. The climb must challenge him to push his limits, but at the same time it needs to be planned so that each move is predictable, rehearsed. He must make just the right move, without even registering an intention to do so, and sense feedback from his body and the rock that will set up the next move. For flow to happen, the demands of the situation must be predictable and success just within reach. It has to be a *stretch goal* or that elusive match between skill and challenge is lost. Honnold won't get into a state of flow during a climb that is super easy for him. Complete control is boring, as our initial rock-paper-scissors study showed. One hundred percent assurance of an outcome robs us of the chance to be the one to make it happen. Boredom and flow, on opposite ends of the control spectrum, are very differently affected by the degree of agency. An increase in agency leads to less boredom and more flow. Total predictability and assurance of an outcome, however, makes us irrelevant, and we disengage.

We need clear goals and unambiguous feedback for flow to arise. Flow simply won't happen if we are uncertain about what we want to achieve or if we cannot gauge how far we are from our goal. To avoid boredom, we must be able to effectively articulate our goals, or at the very least, to choose from among available options the goal we want to pursue. Procrastination is perhaps the most obvious symptom of difficulty in deciding on a goal, one that may contribute to feeling "stuck" in the moment when bored. Indeed, those prone to experiencing boredom are also more likely to procrastinate. In particular, highly boredom-prone people engage in a specific form of procrastination best described as "talking yourself out of it"—a kind of indecision when deciding whether or not to start something.[9] It is the coupling of the desire to engage and the failure to launch that is at the heart of what is so aversive about being bored. The fact that flow is accompanied by clearly delineated goals does little to explain how successful goal setting can be achieved or why it goes wrong when we're bored. However, we do not have to be engaging in goal-directed behavior with clear feedback to avoid boredom. Daydreaming is a case in point. Arguably without a clearly specified goal and lacking in any form of feedback, daydreaming can nonetheless be absorbing.

Flow dissolves when we succumb to distraction. For Honnold or Petit, any lapse in concentration would be catastrophic. Time and time again, research has shown that boredom is accompanied by failures of concentration, whether lapses in everyday tasks (e.g., pouring orange juice on your cereal) or poor performance on laboratory tasks.[10] But are there states in which concentration is not required and yet boredom can be kept at bay? In other words, do you need to concentrate in order to avoid boredom? Relaxing on a beach seems to require very little concentration. And most of us would say it is not boring. Just the same, we can recall times when we struggled to relax amid the sun and sand, restless and wanting to do something else. Relaxation has become boredom. So while concentration is critical to flow, and failures of attention or concentration are common to the experience of boredom, it does not follow that the absence of concentration necessarily leads to boredom. Our minds must be occupied, but they do not need to be intensely focused on error-free performance in order to sidestep boredom.

In a state of flow, all awareness of the self dissipates. Flow, in this sense, is close to the opposite of anxiety. Writ large, anxiety represents our concern for ourselves in the face of threat—real or perceived.[11] When so intensely occupied by an activity that distraction is eliminated, our everyday fears and concerns recede into the background. Here again rock climbers provide a good example. Far from seeking thrills or tempting death, what brings these athletes back time and again to a challenge many of us would see as terrifying is the sense of complete calm that accompanies the climb. Indeed, many of Csikszentmihalyi's respondents highlight the fact that they are seeking out the feeling of flow, and once in the midst of it, they feel no fear, no anxiety whatsoever. Honnold claims, "I generally climb hard routes in the absence of fear."[12] In contrast, boredom is strongly associated with self-awareness. The bored person is painfully aware that they have been unable to lose themselves in activity. With respect to self-focus, boredom is indeed the antithesis of flow. Similar patterns play out for those who often feel bored, with strong associations with self-focus, anxiety, and neuroticism.[13] In this light, boredom and flow represent opposing ends of a continuum of focus on the self.

From his extensive interviews, particularly with those involved in extreme sports like rock climbing, Csikszentmihalyi suggests that for flow to exist, whatever we are engaged in must feel effortless. This does not mean the absence of complex physical or mental skill, just an ability to let those skills flow with ease and grace. Anyone who saw Honnold complete the first free-solo climb of El Capitan would not doubt that substantial physical and mental skill was required.[14] Honnold himself notes, "I'm not thinking about anything when I'm climbing, which is part of the appeal."[15] Clearly the placement of his hands and feet requires precision and a great deal of preparation, skill, and practice. But in execution it *feels* effortless.

For Honnold, rock climbing is its own reward. Being a star of the climbing world no doubt has other perks, but we doubt that is why he climbs. People fortunate enough to experience flow have found activities that are intrinsically rewarding. The activity is pursued purely for the pleasure and reward that comes from being fully immersed and optimally challenged. Here too, flow is the opposite of boredom. When bored, we are unable to find anything that we want to do, let alone something that we want to do simply for its own sake. Intrinsically rewarding activities easily capture and hold our attention; we never have to force ourselves to do them. Intrinsically rewarding activities and boredom are like oil and water—they simply don't mix.

The last of the core components of flow involves distortion of time. Csikszentmihalyi points out that while this is commonly a compression of time—hours feeling as though they pass by in minutes—this is not always the case. Intense attention to detail, to the here and now of an experience, can also make time feel as though it is standing still.

In either case, the person in a state of flow feels liberated from time. The distortion of time associated with flow—whether compression or expansion—is linked with the intensity of concentration that is a key part of the enjoyment of the state. This contrasts with boredom, where time drags on.[16] Our perennial example is waiting at the Department of Motor Vehicles for your number to be called, nothing to do but pass the time. Time that moves at glacial speeds is a com-

Table 9.1. Flow versus boredom

	Flow	Boredom
Skill-challenge balance	"Just right"	Too easy / too hard
Control	Optimal zone	Too much / too little
Goal setting	Clear objectives	Failure to launch
Concentration	Intense focus	Lapses in attention
Sense of self	Self "dissipates"	Self-focused
Effort	Effortless	Effortful
Motivation	Intrinsic	Ineffectual
Time	Distorted (condensed or expanded)	Drags on

ponent of boredom and is also prominent in those who chronically experience it. Time crawls when we're bored because we are not mentally engaged,[17] and the slow march of time is one of the central reasons we find boredom to be so unpleasant.

From Table 9.1 we can see that in many ways boredom and flow do seem to occupy the opposite ends of spectra that describe the two states. But to suggest that flow is the opposite of boredom is not the same thing as saying that boredom can only be averted by entering a state of flow. Flow represents a kind of particularly intense engagement, and such a level of intensity may not be necessary to avoid boredom. What is critical about flow as boredom's opposite is not the intensity of the experience but that it represents successful engagement with the world and successful deployment of our skills and talents.

If flow is not the only avenue of relief from boredom, what else might qualify?

Beyond Flow

Being interested in something is by definition characterized by enhanced engagement, evident in persistence and concentration; and, as such, seems like another plausible opposite to boredom.[18] Clearly, if we feel interest while doing something it would make no sense to say we are also bored. We've argued that something is boring if it does not correspond with our desires and fails to occupy our minds. But what makes something interesting? Are certain objects or activities inherently interesting? Exploring these questions can clarify the ways in which interest is, and is not, the inverse of boredom.

The philosopher Daniel Dennett highlights this problem of what makes something interesting in what he calls Darwin's "strange inversion of logic."[19] It is a counterintuitive claim that reframes how we think about interest and engagement. Honey offers a good example. Dennett's claim is that, contrary to what common sense might suggest, we do not like honey because it is sweet. Instead, honey is sweet because we like it. The logic goes something like this: what is important about honey is its glucose content, and as Dennett points out, endless examination of glucose molecules will tell you nothing about why it's sweet. However, over evolutionary history it has been important for our ancestors to seek sources of glucose for our energy needs. When we found a good source, we associated the taste of that source with pleasure, to reinforce its important function—keeping us alive. Eventually, this sense of "liking honey" is translated as a preference for sweetness. Thus, evolutionary forces shaped us to feel the pleasure of sweetness—to enjoy the feeling of glucose molecules on our tongue—as a way of ensuring we eat honey whenever we find it.

The same inversion of logic can be directed at cute, sexy, and funny things. Babies are not inherently cute but are perceived so because it is adaptive for us to care for them. Certain body shapes are considered sexy because it is adaptive for us to procreate. Unexpected conclusions to jokes are funny because it is adaptive for us to normally engage in effective, logical problem solving—what Dennett calls the "joy of debugging." In other words, detecting faulty reasoning is instructive, and we experience it as funny to ensure we keep doing it.[20] Desirable qualities—sweetness, cuteness, amusement—are not inherent in the things themselves but are the result of evolutionary processes that have shaped our desires over time. This inversion of logic, although strange at first, makes sense when it comes to things like the near-universal desire for calorie-rich food. But how could Darwin's inversion of logic account for one man's abiding passion for traffic cones?

David Morgan, acclaimed Dull Man of Great Britain, is the proud owner of the world's largest traffic cone collection.[21] Remarkably (for those in the know), his collection includes a highly valued 1956 Lynvale rubber cone from Scotland. David enthusiastically describes his passion: "There are so many different shapes, sizes and colours. And the models are always changing." The Dull Men's Club, known for taking pleasure in "everyday, run-of-the-mill stuff," recently applauded David's interest by featuring him in their 2015 calendar of Dull Men. The interests of David and his calendar mates, such as Kevin Beresford of the United Kingdom's Roundabouts Appreciation

Society are surprising and strangely fascinating. Their peculiar interests make it clear that objects or activities are not inherently interesting. Interest, like love, is in the eye of the beholder. But how is it possible that one person can derive pleasure from collecting traffic cones while another must walk across a wire strung between the Twin Towers in New York and another must climb a 3,000-foot cliff face with no ropes?[22] It is difficult if not impossible to understand the vast breadth of human interests by pointing to evolutionary programming—people's interests are simply too idiosyncratic.

However, applying Darwin's inversion of logic to David Morgan, the traffic cone man, we would say that he is not interested in traffic cones because they are interesting, but rather traffic cones are interesting *because* he is interested in them. In this way of thinking, interest is an outcome, not a cause. David presumably came into the world like the rest of us; desiring foods (like honey) that are sweet. We seriously doubt his attraction to traffic cones is similarly innate.[23] If evolutionary processes did not directly shape his unusual hobby, what did? We think it could have been boredom.

Our biology has wired us in such a way that we like being mentally engaged; it feels good, and the opposite feels bad. The good feeling of engagement stems from the fact that it is adaptive for us to engage with the world to gain mastery and develop skills. Our desire to be mentally engaged will therefore push us toward engagement. Once we become engaged with something, magic happens—our attention to that thing *makes it interesting.* Masato Nunoi and Sakiko Yoshikawa from Seisen University and Kyoto University in Japan, demonstrated that things we more deeply pay attention to are preferred compared to things we pay less attention to.[24] They showed people abstract shapes and asked them to do one of two things—report the position of the shape on the screen or tell the experimenters whether they were able to make some association with the shape. They might, for example, say that the shape resembles a dog—an exercise much like finding shapes in the clouds. For some of the shapes people completed this task only once, and for others, five times. Later on, they showed people the same shapes intermingled with some new abstract shapes they had never seen and simply asked them to rate how much they liked the shapes. You might imagine that novelty would win out, and people would rate the new shapes they were seeing for the first time higher than the others. That's not what happened. The shapes people preferred the most were the ones they had seen multiple times and with which they had made frequent associations.

The power of engaged attention to make something interesting is similar in many ways to an older research finding known as the "mere exposure effect," derived from pioneering work by Robert Zajonc from the 1960s onward. Essentially, the mere exposure effect shows that people prefer things they are familiar with. Even if you rate a song on first listening as horrible, your feelings for it will rise on a second listening and can be higher than songs you hear for the first time.[25] We seem hardwired to like familiar things. Robert Zajonc has been quoted as explaining it like this; "If it is familiar, it has not eaten you yet!"[26]

Critically, this preference for familiar things really sticks; in the Nunoi and Yoshikawa study they tested people six weeks after they'd first seen the abstract shapes, and people still preferred what they had more deeply paid attention to and explored. Things, especially things like traffic cones that we are not biologically predisposed to attend to, are difficult to deliberately focus on. But if we can rise to the challenge and devote ourselves to attending to them, they eventually become interesting. Boredom does not push us toward traffic cones, or anything else for that matter. Boredom pushes us away from the uncomfortable feeling of being mentally unengaged. If a traffic cone happens to be in front of us when that push comes, we might develop a lifelong passion for them, as David Morgan did. In that sense, boredom is a pre-interest mechanism that motivates us to become mentally engaged. Once engaged with whatever it may be, interest develops, which in turn deepens and maintains our engagement.[27]

The moral of the story is that you too could have your own peculiar—possibly dull to others—interest if you give some unassuming object enough of your attention and if the push of boredom gets you started. Our argument that boredom is a pre-interest mechanism that pushes us to engage with our environment suggests that boredom and interest are functionally different. But what about the *feeling* of boredom vs. the *feeling* of interest? Are they opposites? We would argue no. Boredom is about our state of mind, namely that we are not mentally occupied. Interest is about something more particular—fly fishing, classical music, even traffic cones—with which we engage. Boredom is a contentless, objectless yearning to be engaged. Although not strict opposites, they never coexist. If we are feeling interest, we can't also feel boredom. More pointedly, boredom is not merely a lack of interest. Boredom involves a restless desire to be engaged, in addition to the sense that what is at hand is not interesting enough to satisfy.

Interest is also not the same as flow. Being interested in something can lead to a flow state, but it does not necessarily do so. For one thing, we can be interested in things that are objectively unpleasant—think of horror movies. Flow, by definition, is felt as pleasant. So you could be interested in a horror movie, but the in-the-moment scares could hardly be seen as pleasant. And what of the sense of control or agency that is critical to flow? Think of the almost ubiquitous early scene in horror movies of a cat jumping out from behind the curtain. We might be able to predict the coming jump scare, but our hearts skip a beat nonetheless. At the mercy of surround sound and musical artistry, we can't help but be startled even if we know it's likely to be a trick. So watching a horror movie can be interesting and engaging without being strictly pleasant or requiring us to feel flow. It is the sense of engagement that allows us to effectively connect with the world and ward off boredom.

If we continue to pursue what we might classify as boredom's opposite in a quest to find the best solutions to the discomfort of boredom, we must consider not just interest but also curiosity. It has been said that "The cure for boredom is curiosity. There is no cure for curiosity."[28] So maybe curiosity is a good candidate for boredom's opposite, and perhaps cultivating curiosity might be a great way to keep boredom at bay. As we hinted at in Chapter 6, humans have always been curious about the world around them, enough to push exploration of inhospitable regions from polar ice caps to outer space. Like interest, curiosity is about feeling drawn to something in the environment. We want to know what is around the next corner. And, as with interest, it is incoherent to say we are both curious about and bored by the same thing. On the other hand, curiosity and boredom do share a similar function—both signal the motivation to explore.

Curiosity spurs exploration of something in particular, boredom pushes us to address the uncomfortable feeling of being mentally unoccupied more generally, and interest helps to maintain engagement once we have started. Exploration itself serves many roles: searching for resources (for example, food, shelter, mates), seeking information to fill explanatory gaps (such as, why are grapes more plentiful on this side of the valley?), learning how the world works (think, for example, of infants repeatedly dropping objects from their high chairs), and so on. Exploratory behavior, spurred on by curiosity, minimizes what behavioral economists refer to as opportunity costs.[29] That is, if we decided we were content with our lot in the world and never explored our environs to discover new things, we would run the risk of missing out on more bountiful resources or opportunities. The berries in our valley may be tasty and sufficient for now, but what if the berries just around the corner are bigger, juicier, and more plentiful? Without exploration, we would never know.[30]

Curiosity is thought to consist of two types: information seeking and stimulus seeking.[31] The former is intended to address knowledge gaps. Where there are holes in what we know of the world, we seek out more information. The latter is related to the need for varied experiences—not thrill seeking per se, but the desire to seek out new sensations and experiences. The two types are clearly related. Any search for new sensory experiences might highlight the things we don't know about the world but would like to: why do certain things we've encountered for the first time taste, feel, or look the way they do?

Researchers have yet to explore the relationship between curiosity and boredom in great detail, but there are some indications that the two are negatively correlated.[32] In an academic setting, not only are curiosity and boredom at the opposite ends of the engagement continuum, they also demonstrate distinct relationships with learning and the value we associate with the task at hand. Much of this might seem obvious. Of course we value something that we are curious about or interested in more highly than something that we deem to be boring. Curiosity and boredom also show opposing influences on learning strategies. When curious, we tend to adopt optimal learning strategies, engaging in rehearsal of learned information, thinking more critically about the information as presented, and using strategies to elaborate or go beyond the surface material. All of these learning strategies are negatively correlated with boredom. When bored, we struggle to adopt an effective attitude toward learning—an attitude that would normally be well cultivated by curiosity.

Curiosity and interest represent states in which we successfully occupy our minds. They illustrate that we can be engaged without necessarily experiencing flow. However, flow, curiosity, and interest all differ from boredom, for all involve effective pursuit of goals. Whether our desire to achieve something leads to that intense experience of the world falling away, to concentration so powerful that hours seems to pass in minutes or seconds is less important as an antidote to boredom than the fact that we have an actionable goal that engages our mind and allows us to express our skills and desires.

But are there other ways we can avoid boredom without needing a specific end in sight?

Idle but Not Bored

Try to recall the last time you really, deeply relaxed. Perhaps it was at the cottage, or on a beach with the latest Jo Nesbø thriller in hand, or even just on the couch, watching the afternoon sunlight slowly shift across the hardwood floor. Wherever it was, your mind was clear; thoughts of work were nowhere to be found; worry and tension were blissfully absent. Nothing needed to change, and you wanted for nothing. That's the key to relaxation—a conspicuous absence of any restless urge to be doing something. No particular goal in sight, no desperate need for productive pursuits, just time to be.

Relaxation is a low-energy, pleasant feeling. Boredom is in part a restless desire to have something to act on. In the throes of that desire we can't relax. And boredom is always unpleasant while relaxation, by definition, is pleasant. But can the contrasts go deeper? At its core, relaxation is the absence of unfulfilled desires. It's this absence of *yearning* in relaxation that qualifies it more profoundly as a candidate for the opposite of, and antidote to, boredom. Boredom is underpinned by a strong feeling that the desire to be mentally engaged is going unfulfilled. When we are relaxed we are free, unburdened by unmet desires.[33] Indeed, simply telling highly boredom-prone people to relax works to decrease their feelings of in-the-moment boredom.[34]

At first blush, it may seem that boredom and relaxation actually share something in common—an underutilized mind. But they are clearly not the same thing. When relaxed, we are unburdened by goal-directed desires, which is not the same as being mentally unoccupied, the precondition for boredom.

Even during states of relaxation our mind is still occupied. Perhaps we are lost in a daydream, planning for the future, or puttering around in the garden. Our mind is engaged, albeit in a somewhat unfocused manner. We may not even be deliberately directing the content of our thoughts but letting our mind drift here and there as it pleases. Nevertheless, our mind is still occupied in a way that is not the case when bored.

In Japan a recent practice known as shinrin-yoku, or "forest-air bathing," has gained in popularity.[35] The practice simply involves spending time in nature to promote health and well-being. Comparing time in the forest with time in an urban setting showed that hostility and depression decreased significantly when people were in the forest. So too did boredom. Maybe there is something particularly important about being in nature; or maybe it is simply a place where the demands of life recede and we can more easily relax. Indeed, many of us find it difficult to relax in our day-to-day lives. Without constant activity, striving for accomplishments or looking for attention-grabbing situations, we often slide into boredom and quickly find ourselves seeking succor from an exciting environment. Relaxation requires the ability to be idle without becoming bored.

It's not necessary to be in a state of flow, interest, curiosity, or relaxation to avoid boredom. But each of them are in some way incompatible with boredom; and they each clarify and deepen our understanding of it. When bored, our mind is unoccupied, and we can't remedy the situation because our desire for engagement is *ineffectual*—we're stuck in the moment with no solutions presenting themselves. We are caught in boredom's desire conundrum. Boredom's opposites, in contrast, are characterized by fulfilled desires, mental engagement, and a strong sense of agency. To return to free solo climber Alex Honnold, we would speculate that his love for the sport encompasses all of boredom's opposites: that at times he may find himself in the state of flow, that he is probably interested in honing his technique, that new climbs with new partners are likely something he is curious about, and that a successful climb is followed by intense satisfaction and possibly a moment to relax. We have no idea if this means he never experiences boredom. But climbing or other all-consuming sports may fulfill our need for agency and reduce the scourge of boredom. What we can say is that if each of these varied forms of engagement are indeed opposites of boredom, perhaps in their pursuit we can prevent boredom from taking hold. But that still leaves us with the question of how we should respond when boredom does strike.

CONCLUSION

• • •

It hits you while gazing out at your front yard that it would have been cool to have made a time-lapse film. Not to chart the changes with the seasons, but to chart the changes wrought by your own hands—time lapse over years, not months.

What was once merely lawn—an assortment of large and small leaf weeds, really—has now been naturalized. Tulips in a dozen different colors signal the beginnings of Spring. To the right, you've planted a tri-color beech, its variegated leaves a deep purple, rimmed in pale pink. To the left are veggie beds, denuded kale stalks, and remnants of bean vines—all that's left from last year's haul. A winding path of uneven width cuts its way through it all, with a weeping mulberry bush surrounded by creeping thyme centered in front of the path. It has taken years to put together.

And it all needs weeding. Again.

Every time you start on the weeds you wonder if you've done nothing more than make endless work for yourself. No doubt you have, yet you like it nonetheless. There is a sense of progress as you wind your way through the yard removing what doesn't belong. There's a goal in sight, and when you reach it, and it all looks great, you know that it's the work of

> your hands. Not prizeworthy, maybe not even praiseworthy, but undoubtedly your success.
>
> By all measures, weeding should be boring. It is monotonous, lacks any serious challenge, and its meaning or value to your life is tenuous. So why do people enjoy weeding? It is engaging, somehow. There is no uncertainty about what you are doing now or what you have to do next. And there is that sense that you have accomplished something. Your actions effected immediate and gratifying change.

• • •

Boredom is a call to action. We can respond to the call in myriad ways. We can try to mask our boredom, to outrun it. Plunging down the rabbit hole of the Internet or social media certainly occupies time, but at some point it dawns on us that there has been little value to what we've been doing. Seeking thrills or constant novelty and excitement is no better. Risky at best, the strategy is simply unsustainable. The belief that we are entitled to and capable of engendering a constantly changing, endlessly stimulating and compelling experience dooms us to continual struggles with boredom. Boredom signals the need to act, to change whatever it is we are doing, to occupy our minds. But to morph that signal into one that then necessitates perpetual exhilaration makes the desire impossible to satiate.

We should neither mask nor try to outrun our boredom. In fact, responding well to the boredom signal gives it value. Accepting the discomfort protects us from the ruin of stagnation precisely because it motivates action. In this sense, being bored is neither good nor bad. The key to unlocking boredom's promise and avoiding its scourge is in our response.

But responding to boredom is no simple task. The bored person wants to do something but at the same time doesn't have anything in mind they want to do—an impossible challenge, a kind of Gordian knot.[1] Amid the distress of boredom we may imagine that a solution lies in excitement, but this is what makes the signal enigmatic; the obvious solutions may be more likely to compound the problem. All too often in the throes of boredom, we expect the world to solve the problem for us. We may haphazardly try different things to see what satisfies. Or, like children imploring their parents to rectify their boredom, we may demand that others in our lives should fix our boredom for us. Just as solving the Gordian knot does not lie in the obvious, so too the solution to boredom cannot come from anything blatantly evident out there, in the world. The solution must come from within us.

We are driven to be the authors of our lives. But, when bored, we have lost our agency, the sense that we are in control. The story of the angler and the cork in Chapter 2 comes to mind. The cork, floating aimlessly in a tumultuous sea, has no agency. It is at the whims of nature. In contrast, the angler can choose where and when to drop anchor or to pack it up and head to shore to outrun a coming storm.[2]

Boredom tells us that we need to reassert our agency, but at the same time it reminds us that our agency is limited. We are neither gods, capable of forcing the world to bend to our will, nor passive receptacles, waiting to be filled and incapable of influence. We occupy an awkward middle ground that can be difficult to accept. But when the messenger of boredom arrives, it is best to take actions to fulfill our agency, limited though it may be. All too often we do the exact opposite; we take actions that curtail our agency and make our struggle with boredom worse in the long run.

Writing in the first half of the twentieth century, Bertrand Russell claimed that boredom is actually becoming less prevalent but more menacing.[3] According to Russell, the menace is related to our increasing fear of boredom. What are we so afraid of? Boredom is distressing, so for that reason alone, we fear it, but at least two other sources come to mind. The first is a fear of our own inadequacies, and the second is a fear of failure. Both are a challenge to agency.

On first glance, it might seem strange that we often respond to boredom out of fear and in counterproductive ways. But it's not really that surprising. When faced with distressing circumstances of any sort, we seek relief. And the more distressed we are, the more preoccupied we become with finding immediate relief. Long-term consequences recede and immediate outcomes loom large.

When afraid of boredom, we desperately seek out activities that are effortless and that we hope will free us from boredom as quickly as possible. Many such activities are *designed* to control our attention, essentially trapping us in mindless engagement. Clickbait on websites keeps us in the crosshairs of advertisers. Perfectly titrated levels in video games propel us to get to the next stage. Bells and whistles on poker machines are designed to keep us playing, even as we lose our life savings.[4] All these things capture our attention. But they do so by treating us as objects to be acted on, bypassing our need to direct the path of our engagement. Such attention-grabbing devices work all too well in the short term; so well, in fact, that they are irresistible when we are desperate to be rid of boredom. In the long run, the more we allow things external to us to solve the problem of boredom, the more our agency atrophies. The more our agency atrophies, the more vulnerable we become to boredom. It's a vicious cycle that only gains momentum, becoming more and more difficult to stop. Fear fuels the cycle and ensures that we will be continuously dogged by boredom. There is actually no need to be afraid. Negative feelings of any sort, like boredom, are not themselves dangerous; rather, they point to danger that needs rectifying. The critical message of boredom is that our agency is diminished, and we need to do something about it. Shooting the messenger doesn't help. It only leaves us ignorant of the fact that we are failing to satisfy the genuine need—not to be *merely occupied,* but to be challenged to exercise our agency in how we engage with the world, to be the fisherman, not the cork.

What would an adaptive response to boredom look like? As discussed in the last chapter, seeking a state of flow, fostering curiosity, even just relaxing, are all avenues to effective engagement. Beyond these options lies another, one that demands we be in the moment. As a response to boredom, it asks us to avoid looking for avenues for escape and instead turn our focus to the here and now. This inward attention allows us to examine more closely what is driving our experience. Through regular practice, we become more present in the moment and more present to ourselves. In other words, we become mindful.

Mindfulness, a form of meditation, fosters the ability to pay attention to our thoughts and feelings without judgment, and it is linked to lower levels of boredom. People with stronger mindfulness skills report feeling bored less often.[5] Even in pure definitional terms, boredom and mindfulness are incompatible—the more bored we are, the less mindful we are.[6] In part, being mindful keeps boredom at bay by making us less emotionally reactive to a boring situation. As with other distressing feelings, the more we fear and attempt to flee from boredom, the more distressing it becomes.[7] Mindfulness meditation may help us break the cycle of reacting to negative feelings with more negative feelings, preventing us from responding to boredom with fear and hostility.[8] In our attempts to outrun boredom we rob ourselves of the chance to learn how to be in the moment and redirect our energies in positive ways. To discover what it is we most deeply want to do, we need to tolerate periods of down time, time not filled by something guiding our thoughts and behavior from the outside. By accepting the risk of being bored we have the chance to find the antidote. Rather than fighting against boredom, accepting a boring situation gives us what we need to be free of it—the chance to identify our desires and goals so that we become engaged on our terms, as agents, committed to a purposeful course of action.

Rather than claiming it is good to be bored, we claim that it is good, on occasion, to be understimulated by the world. As Andy Warhol famously exhorted: "You need to let the little things that would ordinarily bore you suddenly thrill you."[9] It is good to accept less stimulation, to resist the urge to let things outside ourselves drive and control what we do. In other words, we can consciously choose what we do in our lives and the pace at which we do it.[10]

In allowing our engagement to be determined by outside forces that have been engineered to capture and hold our attention, we become alienated from ourselves. In contrast, when the external forces recede, we have the possibility of finding ourselves again; that is, discovering who we are.[11] This is on one hand a blessing, a necessary precondition for embracing our agency and rooting out boredom. On the other hand, taking a good hard look at ourselves is not always pleasant. Joseph Brodsky, a Russian / American poet and essayist, confounded Dartmouth College graduates with his commencement speech, in which he extolled the virtues of boredom, claiming that it "puts your existence into its perspective, the net result of which is precision and humility. You are insignificant because you are finite."[12]

When the cycle of "desire-action-new desire" comes to a standstill, when we are bored, we catch a glimpse of the ultimate futility of our actions in the face of the infinity of time. As we've already said, boredom reminds us that our agency is limited—we are neither gods nor empty vessels, waiting to be filled. What we need to do is come to terms with our own ordinariness. We must be willing to abide in time and brave the ordinary, and be able to do so without succumbing to boredom.[13] Responding effectively to boredom requires embracing your limits. Boredom is a reminder that you are finite and that your actions are ultimately insignificant; yet it requires you to make choices and engage in projects. For Brodsky that is a key, life-affirming lesson of boredom. Your actions are insignificant, but you must act. This is not a pessimistic predicament. On the contrary, it is life itself: "the more finite a thing is, the more it is charged with life, emotions, joys, fears, compassion . . . passion is the privilege of the insignificant."[14]

Because you are finite and subject to the ever-present possibility of boredom, you can experience passion. Nietzsche put it well when he mischievously pondered that "The boredom of God on the seventh day of creation would be a subject for a great poet."[15] Being omnipotent and immortal means everything is possible, making it difficult to value anything in particular. For you, on the other hand, passionate engagement, born of the knowledge that time is short, flows from human agency and is a surefire cure for boredom. It is also when you are at your best.

When the messenger of boredom arrives, it is wise to take a deep breath, kick out any external forces that might control your attention for you, accept your limits, and pursue actions that fulfill your agency. There is no simple, one-size-fits-all set of actions that will do this for you. Boredom can't tell you what you ought to do; nor can we.

In lieu of simple answers, we suggest the following principles. Seek out activities that clarify, rather than obscure, your desires and goals. Pursue goals that give expression to your values—things that matter to you. Do things for their own sake, rather than as a means to avoid something else. Pick activities that enchant your surroundings so you are drawn into ever-deeper connections (think of David Morgan, fascinated by the nuances of traffic cones!). Act so as to express and expand your efficacy. And find activities that engage you as a unique person and express who you are.

Boredom confronts us with the simple yet profound question: What will you do? It demands an answer. There are few more significant questions.

NOTES

REFERENCES

ACKNOWLEDGMENTS

ILLUSTRATION CREDITS

INDEX

NOTES

1. Boredom by Any Other Name

1. The *Oxford English Dictionary* claims the first use of "boredom" was in Charles Dickens's *Bleak House,* originally published in 1852 / 1853 (Dickens, 2003). The dialogue, however, comes from the adaptation by the British Broadcasting Corporation (BBC) in 2005. The closest quotation from the book is: "My Lady Dedlock says she has been 'bored to death'" (21), with "boredom" first introduced like this: "only last Sunday, my Lady, in the desolation of Boredom and the clutch of Giant Despair, almost hated her own maid for being in spirits" (182).
2. Lord Byron penned the satirical epic poem *Don Juan,* in which he used the adjective *bored*: "Society is now one polish'd horde / Form'd of two mighty tribes, the Bores and the Bored (Byron, 1824 [2005]), Canto 13, Verse 95, lines 7–8.
3. Ralph Waldo Emerson wrote of French boredom in 1841: "This *Ennui,* for which we Saxons had no name, this word of France, has got a terrific significance. It shortens life, and bereaves the day of its light" (Emerson, 1971; see discussion in Paliwoda, 2010, p. 16).
4. The experience we call *boredom* has roots in *tedium, ennui,* and *acedia*—sinful laziness or indifference to one's duties toward God. Thus, boredom did not first arrive when the Industrial Revolution gifted Lady Dedlock with generous amounts of free time.
5. Ferrell (2004); Frolova-Walker (2004); Raposa (1999); Spacks (1995); Svendsen (2005); Toohey (2011); Wardley (2012); Winokur (2005).
6. Toohey (2011).

7. Toohey's own translation (brackets added). See Martin et al. (2006) for further discussion of ancient definitions of boredom.
8. Ecclesiastes 1:9 (New International Version). While this description resembles Seneca's complaints of monotony, it could more accurately be seen to represent the view that seeking material things is ultimately meaningless. Later, the narrator claims that even the attainment of wisdom is meaningless in the face of eventual death and that he "hated life." Perhaps this passage represents a lack of spiritual connection, or even depression.
9. How the official eliminated boredom is unclear—that he did so is literally set in stone.
10. Kuhn (1976). Evagrius of Pontus (345–399 AD), a Christian monk or desert father who is said to have developed the first doctrine of acedia, provides the following description: "The eye of the person afflicted with acedia stares at the doors continuously, and his intellect imagines people coming to visit. The door creaks and he jumps up; he hears a sound, and he leans out the window and does not leave it until he gets stiff from sitting there . . . he rubs his eyes and stretches his arms; turning his eyes away from the book, he stares at the wall and again goes back to reading for awhile; leafing through the pages, he looks curiously for the end of texts, he counts the folios and calculates the number of gatherings" (cited in Nault, 2015, p. 29).
11. It's not until the Renaissance that the phrase "noonday demon" was used to refer to melancholia as opposed to something closer to boredom or acedia. The melancholia experienced by monks is even more explicitly associated with too singular a focus on math and science in one's studies! Solomon (2001).
12. Theodor Waitz made significant contributions to the psychological study of feelings with his publication of *Lehrbuch der Psychologie als Naturwissenschaft* (Textbook of Psychology as a Natural Science) in 1849 (Romand, 2015; Teo, 2007).
13. According to Waitz we only become aware of formal feelings like boredom when the anticipated flow of our thoughts is inhibited.
14. Lipps (1906), quoted by Fenichel (1951, p. 349). Lipps was influenced by Waitz and in turn influenced Sigmund Freud.
15. Galton (1885); James (1900); Waitz (1849).
16. According to many existentialists, lacking meaning is at the core of human suffering—including boredom (Maddi, 1967, 1970). Victor Frankl famously argued that finding and fulfilling a sense of life

meaning is a basic human need (Frankl, 1978). When people fail to satisfy this need they are engulfed by an "existential vacuum." Such individuals "lack the awareness of a meaning worth living for. They are haunted by the experience of their inner emptiness, a void within themselves" (Frankl, 1959, p. 128). Frankl asserts that this vacuum manifests "mainly in a state of boredom" (p. 129).

17. "Now the nature of man consists in this, that his will strives, is gratified and strives anew, and so on for ever. Indeed, his happiness and well-being consist simply in the quick transition from desire to satisfaction, and from gratification to a new desire. For the absence of gratification is suffering, and *the empty longing for a new desire is languor, ennui.*" (Schopenhauer, 1995; p. 167, emphasis added). Indeed, Schopenhauer is quoted as saying "The two enemies of human happiness are pain and boredom" (p. 198).
18. As David Kangas (2008, p. 389) characterizes this view, Boredom "is the experience of being unable to flee oneself toward an object (an interest) and so of being 'stuck' to oneself. In boredom, the self cannot be. It is forced to relate to itself without the support of a 'meaning.'"
19. Kierkegaard (1992, p. 232). Kierkegaard used pseudonyms nested within pseudonyms. So to form clear interpretations of his work or to say what *he* thought about boredom is difficult. One way to read *Either/Or* is to see Kierkegaard laying out existential conundrums for us to wrestle with, without attempting to solve them. In chapter 7, "The Seducer's Diary," the narrator Johannes Climacus spends his time pursuing women. For Johannes, it was the chase, the seduction that was thrilling. Once consummated, a relationship became boring.
20. The full quotation is: "What wonder, then, that the world is regressing, that evil is gaining ground more and more, since boredom is on the increase and boredom is a root of all evil. We can trace this from the very beginning of the world. The gods were bored so they created man. Adam was bored because he was alone, so Eve was created. From that time boredom entered the world and grew in exact proportion to the growth of population. Adam was bored alone, then Adam and Eve were bored in union, then Adam and Eve and Cain and Abel were bored *en famille,* then the population increased and the peoples were bored en masse. To divert themselves they conceived the idea of building a tower so high it reached the sky. The very idea is as boring as the tower was high, and a terrible proof of how boredom had gained the upper hand. Then the nations were scattered over the

earth, just as people now travel abroad, but they continue to be bored. And think of the consequences of this boredom!" (Kierkegaard, 1992, pp. 227–228).

21. A common theme in existential thought is the notion that boredom is both an obstacle and, for the strong of heart and mind, an aid (McDonald, 2009).
22. Ciocan (2010).
23. We showed that time slows down for highly boredom-prone individuals (Danckert & Allman, 2005).
24. What would our experience be like if the world were not this way, if everything was novel and full of meaning? Might we be equally paralyzed, unable to act in a constantly changing world?
25. Psychoanalysis has long focused on libidinal desires—desires we know are socially unacceptable and so repress. Boredom from this perspective is the consequence of this repressive mechanism. Indeed, the article by Greenson (1953) suggests that highly boredom-prone individuals have strong oral fixations.
26. Greenson (1953).
27. Greenson (1953), pp. 19–20.
28. Lewinsky (1943); Wangh (1975).
29. Bernstein (1975) put it this way: "the inability to experience one's own feelings directly and intensely is the root of chronic boredom" (p. 518).
30. Phillips (1994).
31. Tolstoy (1899, part 5, chapter 8).
32. Fromm (1955). The full quotation is: "Man is the only animal that can be *bored,* that can feel evicted from paradise. Man is the only animal who finds his own existence a problem which he has to solve and from which he cannot escape." (p. 24; emphasis in original). Fromm also wrote: "I am convinced that boredom is one of the greatest tortures. If I were to imagine Hell, it would be the place where you were continually bored" Fromm (1963).
33. Sharpe (2011) studied meerkat pups to determine what benefits play confers. The "practice" theory, that animals play to practice predatory or aggressive behaviors, was not substantiated. (The same null finding is found in many species that engage in play fighting.) Nor did increased play reduce aggression (the social bonding theory). Play does confer benefits in terms of successful breeding and decreased mortality, but we are uncertain as to the mechanism behind this (Bradshaw et al., 2015; Lewis, 2000).

34. Seal tossing is likely to be first about subduing prey and / or loosening skin prior to eating, but it may also represent a kind of orca version of playing with your food!
35. Pellis & Pellis (2009); Potegal & Einon (1989).
36. Wemelsfelder (1993). Francoise Wemelsfelder, an animal welfare scientist, promotes a distinct view of the mental life of animals. In contrast to behaviorists (i.e., behavior explained entirely through stimulus-response relations), Wemelsfelder argues that some behaviors—boredom among them—represent suffering that goes beyond simple adaptive responses to impoverished environs. In support of this, animals housed in impoverished environments demonstrate decreases in exploratory and interactive behaviors. Sleep periods are lengthened and stereotypical behaviors, such as pacing, are common. Prolonged exposure to impoverished environs can lead to even more extreme behaviors, including, in primates, excessively masturbating, and eating and regurgitating feces; in horses, biting, and in birds, self-mutilating (plucking their feathers out).
37. Bolhuis et al. (2006); Carlstead (1996); Stevenson (1983). Donald O. Hebb, a Canadian scientist, is often credited with the initial discovery that rats reared in enriched environments showed learning benefits (Hebb, 1980).
38. Burn (2017); Wemelsfelder (1985, 1990, 1993, 2005).
39. Meagher & Mason (2012); Meagher et al. (2017).
40. Rizvi et al. (2016).
41. For current theories in humans, see Hunt & Hayden (2017). For a review of work in humans and nonhuman primates, see Sirigu & Duhamel (2016).
42. Dal Mas & Wittmann (2017); Danckert & Merrifield (2018); Danckert & Isacescu (2017); Jiang et al. (2009); Mathiak et al. (2013); Tabatabaie et al. (2014); Ulrich et al. (2015, 2016); see Rafaelli et al. (2018) for review.
43. Craig (2009); Uddin (2015).
44. Bench & Lench (2013); Elpidorou (2014); Goetz et al. (2014); van Tilburg & Igou (2011); Westgate & Wilson (2018).
45. Eastwood et al. (2012); Fahlman et al. (2013).
46. Frijda (2005) noted that feelings arise from the "functioning of one's information processes" (p. 483), and "pleasure and pain . . . stem from thinking proceeding smoothly or unsuccessfully" (p. 481).

47. Hamilton et al. (1984). "Both intrinsic interest and boredom are 'affects' associated with paying attention. In fact, the extremes of absorbing interest vs. boredom define an affective-experiential continuum which accompanies the cognitive, information-processing act in attention" (p. 184).
48. See Scherer (2005) for a discussion of how to distinguish an emotion from neighboring affective phenomena. There are many approaches to classifying boredom, whether as an emotion, a mood, or even a drive. We define boredom as a *feeling of thinking* (Eastwood & Gorelik, 2019) but acknowledge there are other approaches that are coherent and useful.
49. We do not favor claims such as "boredom is the experience of monotony or under-stimulation." Instead, we claim that there is only one kind of boredom with multiple causes and define boredom in terms of a feeling and the psychological mechanisms that underlie it.
50. Researchers have made use of various physiological metrics to measure engagement in real time. Armed with an "engagement index," it is possible to set up a feedback loop where the task is made more difficult when engagement wanes and made easier when engagement maxes out. Such automated adaptive systems result in optimal levels of engagement and improved performance. Importantly, the EEG components used to determine levels of engagement overlap with the components associated with self-reported boredom (Freeman et al., 2004; Raffaeli et al., 2018).
51. Fiske & Taylor (1984); Stanovich (2011).
52. One of the virtues of defining boredom as a feeling of thinking arising from a desire conundrum and an unoccupied mind is that it accounts for the fact that these four factors—slow passage of time, difficulty concentrating, a sense of purposelessness, and oscillating arousal—often co-occur with boredom, bridging the gap between naturalistic and humanistic accounts. That is, although not necessary or sufficient, a desire conundrum and an unoccupied mind can cause these four factors (Eastwood et al., 2012).
53. Proposed typologies of boredom typically revolve around presumed causes (reactive vs. endogenous), duration (transient vs. chronic), focus (bored with something specific—emotion—vs. general boredom—mood) and degree of pathology (normal vs. pathological). These factors no doubt overlap. One typology suggests there are two kinds of boredom—"situative" and "existential." Situative boredom is

characterized as reactive, transient, focused, and normal, whereas existential boredom is endogenous, chronic, unfocused, and pathological (Svendsen, 2005). In our view, some instances of "existential boredom" would be better characterized as "lack of life meaning and purpose" rather than boredom per se. Despite the fact that boredom is complex and multifaceted, we don't find such typologies helpful. In our estimation, the scientific study of boredom has been hampered by a lack of precise and consistent definitions, as well as an overly inclusive use of the term boredom.

2. A Goldilocks World

1. The original association between boredom and tip-of-the-tongue phenomena comes from the psychoanalytic writer Otto Fenichel (1953). The metaphor of trying on clothing to see what fits comes from psychoanalytic writers as well.
2. We are making a conceptual argument here about how to best define boredom. Our approach can account for boredom in both constrained and unconstrained circumstances and explain how external constraint, along with a wide range of other causes, all result in the same state (i.e., because they all give rise to the same underlying mechanisms). This synthesizes naturalistic and humanistic approaches to boredom in a unified model. Naturalistic approaches emphasize external causes, such as constraint and understimulation, whereas humanistic approaches emphasize internal causes, such as a lack of emotional awareness and meaning. We show how the desire conundrum and an unoccupied mind draw these diverse causal models together. It is worth noting that boredom typically co-occurs with other negative feelings, most commonly (in order of prominence), loneliness, anger, sadness, worry, and frustration. (Being bored increased the likelihood of also being frustrated by 67 percent.) At the very least, this association demonstrates that, based on self-report, boredom and frustration are distinct but sometimes co-occur. As an aside, this same study showed that being bored decreased the likelihood of also being indifferent—bolstering the notion that boredom involves a desire for desire (Chin et al., 2017).
3. Schopenhauer (1995).
4. The harder you try to find the word or name on the tip of your tongue, the more difficult it is to find and the more likely you are to

experience tip-of-the-tongue for the same word or name in the future (Warriner & Humphreys, 2008).

5. Hesse (1951), p. 140.
6. We rarely stare at one thing for long enough to make these visual adaptation effects work. The neurons responding to the image (the Canadian flag) become saturated, and their response rates diminish. When you then look at the blank space, neurons that respond to the opposite stimulus (the opposing black and white regions) fire more than the neurons saturated to the original image, and you get the after-effect.
7. Davies (1926); McIvor (1987a, 1987b); Wyatt & Fraser (1929); Wyatt & Langdon (1937).
8. Cited in McIvor (1987b), p. 179. The full quotation is: "It is true that the pleasure of the craftsman is being crushed by the steady increase in mechanised processes, the result of which is seen in the tendency to rise of sickness rates for 'nervous disabilities' . . . Repetition processes undoubtedly create a weariness not expressed in physical terms but in a desire by the worker for temporary relief from the enforced boredom of occupation in which the mind is left partially or entirely unoccupied. This fact must be recognised for the understanding of sickness records and absenteeism in the industrial population. *Vastly more days are lost from vague, ill-defined, but no doubt very real, disability due to ennui than from all the recognised industrial diseases together.* . . . The uninterested worker is an industrial invalid. Interest in work leads to industrial good health" (emphasis added).
9. In an early review on workplace monotony, Davies (1926) summarized: "The worker feeding a dial machine, or counting screws, is able to perform these operations with a minimum of attention after a time. It is not generally possible to concentrate on anything else, however, without detriment to the work, or even sometimes danger, and there are consequently voluntary restrictions on freedom of movement, necessities of behaviour, and inhibitions of normal impulses (such as looking up at a sudden noise), which have to be preserved" (p. 474).
10. Münsterberg (1913), p. 196.
11. Münsterberg (1913), p. 197. In 1926, summarizing workplace research, Davies concluded: "Boredom is not felt, or at least has no chance of developing into serious nervous exhaustion, if the worker feels that, though the work is dull, it is worth doing (p. 475)."

12. Nett et al. (2010, 2011); Sansone et al. (1992).
13. Barmack (1937, 1938, 1939). Similarly, Fenichel (1951) said boredom "arises when we must not do what we want to do, or must do what we do not want to do" (p. 359).
14. O'Hanlon (1981).
15. Pribram & McGuinness (1975).
16. O'Hanlon (1981); Weinberg & Brumback (1990).
17. Danckert et al. (2018a); Lowenstein & Loewenfeld (1951, 1952); O'Hanlon (1981).
18. Scerbo (1998).
19. Homer, *The Odyssey* (1962 / 1990).
20. Klapp (1986).
21. Struk et al. (2015).
22. Fahlman et al. (2009, 2013).
23. Vodanovich (2003); Vodanovich & Watt (2016).
24. It's not possible to claim that any given individual trait *causes* boredom. The issue is an experimental one; we can't manipulate traits, but rather we can only look at how different traits *correlate* with behavior. And as all first-year psychology students are taught, correlation does not equal causation.
25. Bernstein (1975).
26. Psychoanalytic explanations for the failure to accurately identify emotions would suggest it arises from a stalemate between internal forces. On the one hand, we want to gratify desires—many of which are disruptive or unacceptable. On the other hand, we want to avoid embarrassment or punishment. Our socialization to "do the right thing" keeps disruptive, socially inappropriate desires from eventuating. But we still *feel* the urge to act, despite having no clear sense of what the original desire was, having banished it to the dungeons of the unconscious. Anything we can think to do fails to satisfy the urge because it is too far removed from the banished desire (Lewinsky, 1943; Wangh, 1979).
27. White (1998).
28. Bond et al. (2011); Hayes et al. (2004).
29. Eastwood et al. (2007); Harris (2000); Mercer-Lynn et al. (2013a, 2013b)
30. Hamilton (1981); O'Hanlon (1981); Smith (1981); Zuckerman (1979).
31. Kenah et al. (2018); Kreutzer et al. (2001); Oddy et al. (1978); Seel & Kreutzer (2003).

32. Goldberg & Danckert (2013).
33. On attention deficit / hyperactivity disorder, see Diamond (2005); Matthies et al. (2012); on schizophrenia, see Gerritsen et al. (2015); Steele et al. (2013); Todman (2003).
34. Gerritsen et al. (2014); Hunter & Eastwood (2018); Kass et al. (2003, 2001); Malkovsky et al. (2012); Martin et al. (2006); Wallace et al. (2002, 2003).
35. Carriere et al. (2008); Cheyne et al. (2006).
36. Mercer-Lynn et al. (2013b).
37. Mercer-Lynn et al. (2013a, 2013b, 2014).
38. Deci & Ryan (1985, 2008); Ryan & Deci (2000).
39. Barnett & Klitzing (2006); Caldwell et al. (1999); Weissinger et al. (1992).
40. Sulea et al. (2015); Tze et al. (2014).
41. Isacescu & Danckert (2018); Isacescu et al. (2017); Struk et al. (2016).
42. Isacescu et al. (2017).
43. Action / state orientations are types of self-direction that determine how effectively people formulate and achieve goals (Kuhl, 1981, 1985, 1994). People high in action orientation plan, start, and follow through on goals. People high in state orientation struggle to achieve goals because they are too preoccupied with the current situation, hesitate to make change, and are distracted. Highly boredom-prone people are high in state orientation and low in action orientation (Blunt & Pychyl, 1998).
44. These two styles refer to locomotion and assessment regulatory modes (Kruglanski et al., 2000). Locomotors prefer to "just do it," getting on with things, and they are less prone to boredom. Assessors prefer to "do the right thing," and they are more prone to boredom (Mugon et al., 2018).
45. To say the angler is an agent does not entail a commitment on our part to any notion of free will or the causal sufficiency of desires. How she decided to go to shore is a murky, hotly debated philosophical question. We are merely saying that she exhibits actions based on choices—go west vs. go east (whether or not those choices are fundamentally free, or causally effective)—whereas the cork's movements are not mediated by subjective choice but by forces beyond its control.
46. Although not a focus of the book, we acknowledge structural and systemic forces that hinder agentic engagement and foster boredom.

Hence, the presence of boredom is a social and moral problem as well as the personal call to action we are emphasizing. We are sensitive to the fact that a solely personal account of boredom that is not balanced with an analysis of context risks blaming the victim and reinforcing oppressive structures. We applaud Elpidorou's (2017) call for research on the moral dimensions of boredom and encourage such analyses from different perspectives.

3. The Motivation to Change

1. Chicago held the World's Fair, titled "A Century of Progress," in 1933. Arthur Plumhoff, a real circus performer who did pierce his body with pins and needles and hang weights on them, likely did not perform there. There were, however, exotic animals on display, babies in incubators, and cities of "midgets"—a world's fair we might not get away with today! Plumhoff was also certainly not the first performer (and he won't be the last) to engage in what looked like painful and self-injurious behaviors for a crowd. Mirin Dajo was a stage performer in the mid-1940s who claimed to have been trained by mystics to tolerate exceptionally high levels of pain. His show involved an assistant plunging fencing foils through his body. One website suggests a plausible mechanism—the gradual creation of fistulas, or tunnels of scar tissue, much like those created when you pierce your ears (or tongue or eyebrow or whatever else you choose), that allowed the foils to pass through. Dajo died when a long needle he had swallowed ruptured an artery. (http://www.skepticblog.org/2010/05/13/the-mysterious-case-of-mirin-dajo-the-human-pincushion/).
2. Dearborn (1932). A case report from 1991 (Protheroe, 1991) highlights the challenges faced by people who suffer from congenital analgesia—an insensitivity to pain. A rare, inherited disorder, the patient suffers from self-inflicted wounds.
3. Eccleston & Crombez (1999).
4. Self-control and self-regulation broadly refer to processes by which people bring their thoughts, feelings, and actions into alignment with goals. The "affect-alarm" model of Inzlicht and Legault (2014) suggests that psychological distress represents a conflict state that triggers self-regulatory mechanisms.
5. Sir Frances Galton first wrote about fidgeting as an indicator of boredom in 1885. He measured the "sway" of audience members in a

scientific talk. When the audience was enraptured they held themselves "rigidly in the best position for seeing and hearing." But when bored they would "cease to forget themselves and they begin to pay much attention to the discomforts attendant on sitting long in the same position. They sway from side to side" (pp. 174–175).

6. Many have conflated boredom and apathy. While boredom is correlated with apathy and anhedonia (a loss of pleasure), boredom is a distinct feeling (Goldberg et al., 2011). Similarly, van Tilburg and Igou (2017) have shown that boredom can be distinguished from a broad range of negative affective states, including sadness, anger, frustration, fear, disgust, depression, guilt, shame, and regret.
7. Iso-Ahola & Crowley (1991); Joireman et al. (2003); Kass & Vodanovich (1990); Mercer & Eastwood (2010).
8. Both state and trait boredom may manifest differently in different cultures. The largest distinction made in this sense is between individualist (e.g., Canada, the United States) and collectivist (China, India) cultures (Ng et al., 2015). While we acknowledge cultural factors will play a role in the expression of boredom, there is currently too little research for us to go into depth on the issue.
9. Elpidorou (2014). Andreas Elpidorou should also be credited with introducing us to the pain metaphor at a workshop on boredom and mind-wandering held at the University of Waterloo, October 2015.
10. Trait boredom is commonly referred to as "boredom proneness," probably due to the most commonly used measure—the Boredom Proneness Scale (BPS; Farmer & Sundberg, 1986). We developed a shorter version of the scale (Struk et al., 2017) that addresses some of its shortcomings (Melton & Schulenberg, 2009; Vodanovich et al., 2005). Our version has a single-factor structure that addresses the need for engagement.
11. Elpidorou (2014); Bench & Lench (2013); van Tilburg & Igou (2012).
12. The (probably) apocryphal story of Humphrey Potter was taken from Adam Smith's book *An Inquiry Into the Nature and Causes of the Wealth of Nations* (p. 14). To our knowledge, the first person linking Humphrey's story to boredom and artificial intelligence was Jacques Pitrat, the AI researcher discussed later in this chapter (http://bootstrappingartificialintelligence.fr/WordPress3/2014/05/when-artificial-beings-might-get-bored/).

13. Elsie Nicks is a real historical figure (Cairns et al., 1941), although our portrayal of her may not be historically accurate. We present her here as a prototypical patient with akinetic mutism.
14. Now known as von Economo's disease after the Austrian psychiatrist and neurologist Constantin von Economo, who detailed the disorder in 1917, encephalitis lethargica is thought to have killed almost 1 million people in the early twentieth century. Also referred to as "sleepy sickness," there is still no known cure, although in one cohort, levodopa, a drug typically used to treat Parkinson's disease, led to brief periods of relief from symptoms—and formed the inspiration for *Awakenings*, a book by Oliver Sacks that was subsequently made into a movie.
15. In the original description of Elsie's case there was a hint that she still had goals and desires but was unable to act on them. When a chocolate was placed in her hand, she attempted to get it to her mouth, but when she dropped it, she made no attempt to retrieve it. She even made "tentative chewing movements," again suggesting she wanted the treat but couldn't will herself to get it (Cairns et al., 1941).
16. Marin & Wilkosz (2005); Mega & Cohenour (1997); Németh (1988).
17. A Pyrrhic victory is one that comes at a huge cost. The phrase is derived from King Pyrrhus of Epirus, who defeated the Romans at Heraclea and Asculum around 280 BCE but suffered massive losses. While the Romans also lost many men, they could withstand their losses more effectively. Plutarch wrote that Pyrrhus remarked: "If we are victorious in one more battle with the Romans, we shall be utterly ruined" (http://penelope.uchicago.edu/Thayer/e/roman/texts/plutarch/lives/pyrrhus*.html).
18. We do not intend a criticism of functionalism. However, we do believe there is virtue in anchoring the definition of boredom in subjective feeling, and that more work is required to carefully analyze functional states in animals and machines that might mirror the functional state of boredom in humans.
19. Breazeal (2009). To watch Kismet in action, see: https://www.youtube.com/watch?v=8KRZX5KL4fA.
20. Turing (1950). This is the paper in which Turing introduces the "imitation game," in which a person must determine whether natural conversation responses represent another human or a machine. The full quotation is: "Instead of trying to produce a programme to simulate the adult mind, why not rather try to produce one which simulates the child's?" (p. 456).

21. Pitrat (2009).
22. We can't say anything about the conscious life of a bored AI system. Perhaps CAIA and Kismet are not *consciously* bored, but we claim they experience something that functions similarly. That is, boredom as a motivational state pushes us to seek satisfying engagement. The AI systems discussed here experience the same push.
23. In Chapter 7 we deal with the relation between boredom and meaning in depth, and in Chapter 8 we delve into this notion of boredom at the extremes of monotony versus random noise, which originates from Orin Klapp's 1986 book *Overload and Boredom*.
24. Burda et al. (2018).
25. Yu et al. (2018).
26. Godin (2007). The axiom "winners never quit and quitters never win" is attributed to American football legend Vince Lombardi, the man the Super Bowl trophy is named after. Lombardi's own coaching record is impressive, never having had a losing season and amassing a 90 percent win rate in postseason games. As a maxim delivered to players and sports fans this may seem inspirational, but as Godin argues, clearly there are circumstances where quitting makes sense. A recent story illustrates the point. A couple sailing from New Zealand to Australia as part of their ten-year plan to circumnavigate the globe found themselves in rough seas. Their rudder was disabled, their boat had been knocked down twice and had capsized once in waves the two described as "the size of buildings." They abandoned ship after activating their emergency signal and were rescued by a container ship. Clearly, sticking with their plan to sail to Australia could have been fatal (http://www.abc.net.au/news/2017-03-30/yacht-abandonded-after-rescue-recovered-off-eden/8402048).
27. White (1959).

4. Across the Life Span

1. This is a fictional adaptation of an actual case reported in the *Telegraph* on 28 January 2014, "Great-grandmother, 76, shoplifted because she was 'bored of being old.'" http://www.telegraph.co.uk/news/uknews/law-and-order/10601150/Great-grandmother-76-shoplifted-because-she-was-bored-of-being-old.html

2. This is the problem of WEIRD science (Heinrich et al., 2010)—science that has a unique sample of Western, Educated, Industrialized, Rich, and Democratic subjects. How confident can we be in generalizing our findings when the samples we test are so narrowly defined?
3. Giambra et al. (1992).
4. This is known as the age of majority, a legal fiction as opposed to a biological or psychological reality. The age of majority describes the point at which society decides that minors become adults and assume certain rights and responsibilities, e.g., https://en.wikipedia.org/wiki/Age_of_majority. See also https://en.wikipedia.org/wiki/Age_of_consent for variations in the age of consent—again, highlighting the distinction between biological and legal milestones.
5. Riem et al. (2014).
6. Phillips (1994).
7. Lehr & Todman (2009).
8. Fogelman (1976). This study used data from a National Child Development Study. Parents of bored kids claimed that their minimal leisure activities were not for want of options. Like the Lehr and Todman study, higher boredom levels were associated with poorer performance at school and lower socioeconomic status.
9. Russo et al. (1991, 1993).
10. A recent natural experiment in New Zealand highlights how adding constraints (or removing them) might influence children's behavior. One school removed previous restrictive playground rules (e.g., no scooter riding on the playground). Teachers found that with *unrestricted* play came increases in imagination and learning and decreases in bullying. Lifting rules and regulations on the playground may also go a long way toward eliminating boredom. (http://nationalpost.com/news/when-one-new-zealand-school-tossed-its-playground-rules-and-let-students-risk-injury-the-results-surprised).
11. Steinberg (2005); Piaget (1999).
12. The last of Piaget's (1999) developmental stages is the *formal operational stage,* characterized by an increased capacity for abstract reasoning, problem-solving, and logical reasoning.
13. Steinberg (2005).
14. Dahl (2001, 2004). The quotation—actually, a subtitle—is on page 17 of the 2004 article.
15. Harden & Tucker-Drob (2011).
16. Spaeth et al. (2015).

17. Harris (2000). In describing the state of boredom, restlessness was reported 26 percent of the time and wandering attention 22 percent of the time. As we argue elsewhere, the common association between boredom and restlessness suggests it is subjectively felt as a high arousal experience (Danckert et al., 2018a, b; see also Merrifield & Danckert, 2014). In the learning environment boredom may also be a kind of protest. For teens, willfully opposing what is available operates as an assertion of their independence and may trump the need to be mentally engaged. Being bored, then, is a way of rejecting the adult world.
18. Caldwell et al. (1992).
19. Haller et al. (2013).
20. Work on leisure boredom implies that the constrained environment of schools can't be the only culprit in teen boredom. In one study (Larson & Richards, 1991), results showed that boredom was consistent across the contexts of school, home, and leisure time.
21. Miller et al. (2014). By "more sexually aggressive," the authors mean that male youth who were more bored in grade 9 claimed, in grade 10, that they would not cease a sexual act if their partner asked them to.
22. Willging et al. (2014); see also Wegner & Flisher (2009). For evidence boredom is higher in rural settings, see Patterson et al. (2000).
23. Sharp & Caldwell (2005).
24. We have shown that age is a significant negative predictor of boredom—the older we get the less prone to boredom we become. This relationship held true even when we looked only at people aged 17 to 22 years (Gerritsen et al., 2015; Isacescu et al., 2017).
25. As was the case with the transition to teenage years, the transition to adulthood is associated less with a chronological age (i.e., age of majority) and more with biological maturation of frontal cortices. Interestingly, in Portugal people are precluded from running for public office until the age of 25—presumably when society feels they have sufficiently matured!
26. This notion of the frontal CEO in part informed the implementation of frontal lobotomies in the 1940s and 1950s. The theory was that a dysfunctional CEO—the frontal cortex—could be disconnected from the rest of the brain, allowing it to go about its business unhindered. The true outcomes of these surgeries were far less benign (Gross & Schäfer, 2011).
27. Each of these executive functions is distinct, making the umbrella term of executive function somewhat controversial. One model of

executive function, the "unity and diversity" model, suggests that a diverse cluster of functions (e.g., working memory, inhibitory control, abstract reasoning) represent distinct cognitive mechanisms relying on separable but overlapping neural networks (the diversity part). Ultimately, these distinct mechanisms work together to enable our most sophisticated behaviors (the unity part; Miyake et al., 2000).

28. Taylor et al. (2017).
29. The anterior (frontmost) temporal cortex performs many complex functions including semantic conceptualization, and the formation of new memories. The orbitofrontal cortex is important for processing olfactory stimuli (many TBI patients experience *anosmia*—an inability to discriminate smells), as well as representing costs and rewards associated with actions.
30. Alan Baddeley (1996) coined the term "dysexecutive syndrome" as a description of dysfunction in another term he pioneered, the "central executive." It is the central executive that "controls" other subsystems important for memory and cognition. Well before Baddeley, Russian neuropsychologist Alexander Luria brought focus to the frontal lobes in his book *The Working Brain* (1973). Luria notes that the frontal cortex is important for the "control of the most complex forms of man's goal-linked activity" (1973, p. 188).
31. Fleming et al. (2012).
32. We tested thirty-five patients who had suffered moderate to severe TBI and a group of 340 individuals who reported having had concussions. Boredom proneness scores were highest in the TBI group but were also higher in the concussed group compared with healthy people (Isacescu & Danckert, 2018).
33. Chin et al. (2017).
34. This has been a substantial problem for researchers interested in the effects of aging. We can study captive populations, such as infants in day care centers, children in schools, undergraduate students, and we can study populations with more free time, such as those of us in retirement. What this led to for researchers studying aging was the notion that there is a cliff at age 60 over which cognitive skills decline. With more data from people in their forties and fifties, we learned that this is a much more gradual decline.
35. Conroy et al. (2010).
36. Cognitive reserve refers to functional resilience in the face of brain damage or decline. Greater cognitive reserve is thought to be a protective

factor against diseases such as Alzheimer's (Medaglia et al., 2017; Valenzuela & Sachdev, 2006).

37. Best & Miller (2010); see also DeCarli et al. (2005) and Scuteri et al. (2005).
38. Ice (2002); see also Korzenny & Neuendorf (1980). The latter study showed that the elderly watched television for two reasons—information and fantasy, with fantasy engaged as a remedy against monotony and boredom.
39. Shuman-Peretsky et al. (2017).

5. A Consequential Experience

1. For a full description of the circumstances around Flight #188 see the following editorial from *Salon*: https://www.salon.com/2009/12/11/askthepilot344/. Our account is fictionalized but remains largely faithful to the information we could find about this incident.
2. Britton & Shipley (2010).
3. The one exception to our inclusion rule is gambling, which we include because it is such a common *purported* consequence of boredom.
4. Berlyne (1960); Kahneman (1973).
5. Hitchcock et al. (1999); Hunter & Eastwood (2019); Mackworth (1948); Pattyn et al. (2008); Scerbo (1998); Thackray et al. (1977).
6. Kass et al. (2001); Scerbo (1998); Thackray et al. (1977).
7. Scerbo (1998).
8. Wilson et al. (2014).
9. It is important to note that some were ambivalent and others enjoyed the experience (Fox et al., 2014).
10. Havermans et al. (2015).
11. Nederkoorn et al. (2016).
12. Favazza (1998). The research on suicide and boredom is underdeveloped. In one study (Ben-Zeev et al., 2012) depressed, hospitalized patients reported sadness, tension, and boredom several hours before having suicidal thoughts, and boredom was the most robust predictor of subsequent suicidal thoughts.
13. Claes et al. (2001) found women with eating disorders pulled out their hair when bored. Chapman and Dixon-Gordon (2007) reported that female inmates claimed boredom was the third most common trigger for deliberate self-injurious behavior, after anger and anxiety.

14. Lee et al. (2007); McIntosh et al. (2005); Orcutt (1984); Piko et al. (2007); Wegner (2011); Ziervogel et al. (1997). People who use harmful substances report being bored more often than people who do not use these substances (Biolcati et al., 2018; Boys et al., 2001; Caldwell & Smith, 1995; Iso-Ahola & Crowley, 1991; Smith & Caldwell, 1989).
15. The National Center on Addiction and Substance Abuse (August 2003); Caldwell & Smith (1995).
16. These studies followed students from grades 8 to 11 and found that students who had the highest levels of boredom during free time at the start of the study were the most likely to use substances later on (Sharp et al., 2011). Intriguingly, one study found that the association between boredom and future cigarette use only held for teens that were initially at low risk of smoking (Coffman et al., 2012).
17. Sharp et al. (2011).
18. Krotava & Todman (2014).
19. Weybright et al. (2015).
20. Biolcati et al. (2018); Blaszczynski et al. (1990); Bonnaire et al. (2004); Dickerson et al. (1987); Nower & Blaszczynski (2006); Turner et al. (2006); Carroll & Huxley (1994); Clarke et al. (2007); Coman et al. (1997); Cotte & Latour (2008); Hing & Breen (2001); Hopley et al. (2012); Hopley & Nicki (2010); McNeilly & Burke (2000); Mercer & Eastwood (2010); Trevorrow & Moore (1998); Williams & Hinton (2006); Wood et al. (2007).
21. We know of no studies using an experimental design to confirm that boredom *causes* substance use or gambling. State feelings of boredom and the general tendency to be bored both co-occur with a number of psychological factors that have not been carefully ruled out in most studies. In one study that attempted to do so, there was no relation between boredom proneness and problem alcohol use and gambling when controlling for impulsivity (Mercer-Lynn et al., 2013b). It's possible that it's not boredom proneness but some associated characteristic that is the factor driving substance use and gambling.
22. Leon & Chamberlain (1973); Stickney et al. (1999); Walfish & Brown (2009).
23. Koball et al. (2012).
24. Crockett et al. (2015); Moynihan et al. (2015).
25. Moynihan et al. (2015).
26. Havermans et al. (2015).

27. Moynihan and colleagues' (2015) study used boring or sad movies to induce different groups into those mood states. While watching the movies people could snack on foods that were more or less healthy and more or less exciting. (Cherry tomatoes, it seems, are considered healthy *and* exciting, but crackers are not exciting.) Regardless of the movie watched, people ate the same amount of healthy, unexciting food (e.g., crackers). When it came to less healthy, or healthy but exciting foods (cherry tomatoes), it was the people who watched the boring movie who ate more.
28. Abramson & Stinson (1977).
29. Meagher & Mason (2012).
30. Gill et al. (2014).
31. Dahlen et al. (2004); Gerritsen et al. (2014); Leong & Schneller (1993); Mercer-Lynn et al. (2013b); Moynihan et al. (2017); Watt & Vodanovich (1992).
32. Moynihan et al. (2017).
33. Matthies et al. (2012).
34. Pettiford et al. (2007); Witte & Donahue (2000). It is worth pointing out that this finding relates to trait boredom proneness, unlike the study about risky decision making, which looked at the impact of feeling bored in the moment.
35. This quotation has commonly—but not definitively—been associated with the American philosopher Paul Tillich.
36. Boyle et al. (1993).
37. Wink & Donahue (1995, 1997); Zondag (2013).
38. Dahlen et al. (2004); Isacescu et al. (2017); Isacescu & Danckert (2018); Joireman et al. (2003); Mercer-Lynn et al. (2013b); Rupp & Vodanovich (1997); Vodanovich et al. (1991); Zuckerman (1993).
39. Quay (1965).
40. For example, Farnworth (1998) describes boredom in young offenders and touches on boredom as a driver of vandalism. There is a great deal of research on vandalism. There are some theoretical suggestions that it is tied to boredom, but little actual research addressing the link.
41. Spaeth et al. (2015). The researchers found no evidence to support the idea that boredom is an expression of a personality-based, angry, willful rejection of society. See also Caldwell & Smith (2006).
42. van Tilburg & Igou (2011).

43. For a full report of the Christopher Lane story, see: http://www.cbsnews.com/news/christopher-lane-australian-baseball-player-killed-by-bored-okla-teens-police-say/.
44. For a full report of this story, see: https://www.cnn.com/2018/01/23/europe/german-nurse-charged-97-murders-intl/index.html.
45. Greenberg et al. (2004).
46. Weissinger (1995).
47. Fahlman et al. (2009); Goldberg et al. (2011); Mercer-Lynn et al. (2013b); Todman (2013); Vodanovich & Watt (2016). We know boredom and depression are not connected because of other factors like the tendency to experience negative emotions, sensitivity to emotionally upsetting threats, apathy, lack of pleasure, and poor emotional awareness (Goldberg et al., 2011, Mercer-Lynn et al., 2013b).
48. Self-report tools used by researchers distinguish between the tendency to experience boredom, the state of boredom itself, and depression, thereby never confusing a bored person for a depressed person (Fahlman et al., 2009, 2013; Goldberg et al., 2011).
49. Spaeth et al. (2015).
50. Fahlman et al. (2009).
51. Bargdill (2000).
52. Fahlman et al. (2009).
53. Gerritsen et al. (2015); Newell et al. (2012); Todman (2003); Todman et al. (2008).
54. Inman et al. (2003); Passik et al. (2003); Theobald et al. (2003).
55. Mann & Cadman (2014).
56. Tolinski & Di Perna (2016), p. 218. The "chitlin' circuit" refers to a range of venues in the United States where it was safe for African American musicians, comedians, and entertainers to perform during the era of racial segregation.
57. Larson (1990).
58. Gasper & Middlewood (2014).
59. Mann & Cadman (2014).
60. Conrad (1997).

6. Boredom at the Extremes

1. The work that forms the basis for this fictionalized account comes from Heron (1957). Participants were free to leave the experiment at any point (although many stayed multiple days), and most of the

hallucinatory experiences were less ominous than the one imagined here.

2. The work of Hebb and Heron was only partially published. Hebb advocated for publishing all results but was prevented from doing so by the intelligence agencies that had commissioned the work. Eventually the work was misrepresented as the progenitor of psychological torture (Brown, 2007). The actual intention of the work was to determine whether sensory deprivation combined with "propaganda" could successfully change an individual's beliefs in the supernatural (see Raz, 2013, for a description of sensory deprivation studies beyond the work of Hebb).
3. In a recent book examining the history of exploration in Canada by Adam Shoalts, he outlines accounts from the Vikings to early European explorers and consistently cites curiosity, acquisitiveness, and desire for fame as drivers of human exploration (Shoalts, 2017, pp. 24, 53).
4. Bishop (2004).
5. Cook (1909).
6. Bishop (2004), table 3.
7. Sandal et al. (2006).
8. Palinkas (2003); Palinkas et al. (2000).
9. Shiota et al. (2007).
10. Sandal et al. (2006).
11. The origin of this quotation is difficult to determine. It appeared in a letter from a British cavalry subaltern at the front in World War I and was published in the *Times* on November 4, 1914. It has also been attributed to Edward Arthur Burroughs, who wrote about the First World War in *The Fight for the Future* (1916).
12. Bartone et al. (1998).
13. Arrigo & Bullock (2008).
14. For synonyms for solitary confinement, see https://www.muckrock.com/news/archives/2015/jun/16/solitary-confinement-may-go-different-name-your-st/.
15. Arrigo & Bullock (2008).
16. Smith (2006).
17. For Ashley Smith's case, see: http://nationalpost.com/news/canada/ashley-smith-death-ruled-a-homicide-by-inquest-jury. For Adam Capay's case, see: https://www.theglobeandmail.com/news/national/how-a-tweet-led-to-unlocking-adam-capays-stint-in-solitary/article34756517/.

18. For those interested, the website Solitary Watch (2019) has useful information about the practice in the United States.
19. Christopher Burney was a British lieutenant recruited to work for the French Resistance in World War II. He was captured in 1942 and sent to Fresnes prison where he spent 526 days in solitary confinement before being shipped to Birkenau toward the end of the war. His account is one that highlights the need to establish routines in order to survive the monotony of confinement. When he failed to adhere to his routines (e.g., saving meager portions of rations to eat later in the day), his mood fell. His remarkable story, told in his book *Solitary Confinement,* also highlights the basic human need for social contact and variety of experience (Burney, 1952).
20. Ethnographic research involves the researcher immersing themselves within a specific cultural group and observing behavior as unobtrusively as possible. The author of this study (Bengtsson, 2012) spent substantial periods of time in two institutions. She found that at the second of the two she received more acceptance and participation from inmates when she disclosed that her study was about boredom. The inmates became more willing participants when they felt she understood what their lived experience was like.
21. Bengtsson (2012) refers to "edgework"—the notion that youth are exploring the boundary between order and disorder. The closer one gets to disorder, the more exciting it seems. Edgework and the approach toward disorder can in turn be characterized as challenging the status quo and authority.
22. de Viggiani (2007).
23. Lebedev (1988), quotations on pp. 32, 60, 78.
24. J. Danckert spoke with Hadfield by phone on November 22, 2017, but one can find more details of his own account of time in space through his book *An Astronaut's Guide to Life on Earth* (2013), as well as numerous YouTube videos and media accounts replete with claims that he never gets bored!
25. Hadfield, personal communication with J. Danckert, November 22, 2017.
26. Lebedev (1988), pp. 125, 251.
27. Lebedev (1988), p. 81.
28. At times Lebedev's expressions of monotony and dreariness occur quickly on the heels of his expressions of the majesty and awe of space. Just after having spoken of the majesty of seeing Earth from space he says that things were "quiet and dreary on the station" (Lebedev, 1988,

p. 203). Shortly afterward, while contemplating the challenge of redocking, he says, "Later, when we began the actual training aboard the spaceship, it wasn't even very interesting" (p. 241).

29. Lebedev (1988), p. 78.

7. The Search for Meaning

1. Galton (1885). Sir Francis Galton published a short article in *Nature* in which he describes being at a meeting and observing (and attempting to measure as was his wont) the audience members. We have used this paper to construct a fictionalized account of the meeting. Clearly though, Galton was unengaged by the meeting itself.
2. Barbalet (1999).
3. Svendsen (2005), quotations on pp. 7, 30.
4. Frankl (1959), p.129.
5. Kuhn (1976); see also Healy (1984) and Raposa (1999).
6. van Tilburg & Igou (2012, 2016).
7. Eakman (2011); Fahlman et al. (2009); Kunzendorf & Buker (2008); Weinstein et al. (1995); MacDonald & Holland (2002); McLeod & Vodanovich (1991); Melton & Schulenberg (2007); Tolor & Siegel (1989).
8. McLeod & Vodanovich (1991); Tolor & Siegel (1989).
9. Fahlman et al. (2009).
10. Drob & Bernard (1988).
11. Bargdill (2000).
12. In Chapter 2 we review the variety of other factors that cause boredom—lack of life meaning is only one of those causes. Moreover, lacking life meaning does not *always* cause boredom.
13. Eastwood (unpublished data).
14. As we discussed in Chapter 1 we view a lack of situational meaning as a closely associated feature or consequence of boredom, but not central to the definition.
15. It is possible to be occupied by an activity (and thus not bored) that has no significant meaning. On the other hand, it is possible to do something that does not fully occupy our mind (and thus feel bored) that is deeply meaningful. Recall our earlier examples of binge watching inane TV and telling endless knock-knock jokes to our preschooler. The first is not meaningful but not boring, whereas the second has deep life meaning but can get a little tedious.
16. van Tilburg et al. (2013).

17. van Tilburg & Igou (2017).
18. van Tilburg & Igou (2011).
19. van Tilburg & Igou (2016).
20. Coughlan et al. (2019).
21. Nels F. S. Ferre, as cited in Boehm (2006), p. 160.
22. Kustermans & Ringmar (2011).
23. Svendsen (2005).
24. Gosselin & Schyns (2003).
25. The researchers note that because the jumbled versions were the most open to "being cognitively manipulated for a meaningful outcome . . . the results supports theories that emphasize the need for active structuring of the environment" Landon & Suedfeld (1969), p. 248.
26. Brissett & Snow (1993).
27. This idea comes from sociologist Orin Klapp's book *Overload and Boredom* (1986), which we discuss in Chapter 8.
28. This line is from Shakespeare's *Macbeth,* and is the source for the title of William Faulkner's novel *The Sound and the Fury.*
29. Svendsen (2005), p. 32.
30. Elpidorou (2014), p. 2. See also Elpidorou (2018).

8. An Epidemic in the Making

1. Klapp (1986), pp. 1–2.
2. For Socrates, see in Yunis (2011); James (1900); Kracauer (1995). Kracauer thought, much like Socrates, that in the early part of the twentieth century memory was under threat from advancing technologies! In our own time, psychologists examine what they call "cognitive offloading"—relying on Google Maps to find our way rather than learning the route ourselves, for example (Risko & Gilbert, 2016). Thus, it seems we are perpetually concerned with Socrates' lament on the threat of the written word, albeit in the form of new technologies.
3. Klapp (1986), p. 49.
4. Richard Smith (1981) was possibly the first to hint at this hypothesis that those prone to boredom reside at the extremes of monotony and excess variety or sensation. Some indirect evidence supports the notion: on one hand, high boredom proneness is associated with increased sensation seeking (Zuckerman, 1979). On the other, those same highly boredom-prone individuals struggle to simply "get on with things" (Mugon et al., 2018). This is the boredom conundrum—the highly

boredom-prone individual *knows* they want something novel, meaningful, and satisfying to engage in, but are stuck at the choice phase of action, unable to launch (Danckert et al., 2018b).

5. The full list deserves closer consideration.
    i. Loudness (a clamoring to be heard above noise—Klapp's "ego screaming"),
    ii. Decoding difficulty (resorting to jargon to make sense of information and to exclude others),
    iii. Disconnection / incoherence (the unconnected nature of information evident in TV screens with a talking head in one window, weather forecasts in one inset and traffic cams in another, plus a ticker tape of news stories below),
    iv. Bad complexity (complexity without meaning),
    v. Channel clutter (a simple problem of reaching and exceeding capacity for processing distinct sources of information),
    vi. Lack of feedback making it more difficult to sort signal from noise,
    vii. Stylistic noise (think of ever changing fads),
    viii. Pseudo-information (Klapp anticipated the current era of claims of "fake news"—by analogy, counterfeit money makes us feel rich so long it goes undetected), and
    ix. Sheer overload—dramatic increases in the amount of available information.
6. Klapp (1986), p. 106.
7. This analysis was first presented by Klapp (1986).
8. For more on the knowledge doubling curve see: http://www.industrytap.com/knowledge-doubling-every-12-months-soon-to-be-every-12-hours/3950
9. Bornmann & Mutz (2015). Collectively, the number of scientific articles published since 1660 passed the 50 million mark, with an additional 2.5 million now published annually (http://www.cdnsciencepub.com/blog/21st-century-science-overload.aspx). Just two examples highlight the challenge of the information explosion: the Internet is estimated to contain 5 million terabytes (TB) worth of information, and Google has mapped only around 200 TB or 0.004 percent of the total information available. Mapping the connections of an average human brain is estimated to require several billion petabytes (a petabyte is 10^{15} bytes)!
10. We remember when floppy discs were indeed floppy, and a document like a PhD thesis needed multiple 3¼ (non-floppy) floppy discs for

storage. The same document now barely registers on the average USB memory stick.

11. Claims such as "The Internet ruined my brain" or "The Internet is making us dumber"—a claim explored in depth in Nicholas Carr's book *The Shallows: What the Internet Is Doing to Our Brains* (2011), harken back to Socrates' concern that the written word would ruin us. Such claims are hyperbolic and unsubstantiated (much like the potential of an epidemic of boredom). Nevertheless, with the advent of each new advance, from carving on stone tablets to the Internet, it is worth asking what the positive and negative outcomes may be.
12. For a history of the Internet, see Hafner & Lyon (1988).
13. The initial characterization of Internet addiction by Young and colleagues (Young, 1998; Young & Rogers, 1998) casts it less as a substance abuse problem and more as an impulse control issue. This fits with our description that a failure to meaningfully occupy oneself prompts a turn to the Internet for entertainment—quick, easy, but ultimately unfulfilling. The notion of Internet addiction as a diagnosable mental illness has not made its way into the *Diagnostic and Statistical Manual of Mental Disorders,* indicating there is still some skepticism in the field. Internet Gaming Disorder is mentioned in the DSM-V (the latest version of the manual) as an area warranting further research. Some of the earliest work did suggest that people overusing the Internet exhibited behaviors indicative of tolerance (wanting more and more Internet time in order to be satisfied) and withdrawal (exhibiting feelings of depression and irritability when offline), both classic signs of substance addiction (Scherer, 1997).
14. Bernardi & Pallanti (2009); Nichols & Nicki (2004). These studies highlight comorbidities—other mental illnesses commonly co-occurring with Internet addiction (e.g., depression and impulse control disorders). People in the Bernardi and Pallanti study reported elevated levels of boredom when they were *prevented* from accessing the Internet, another clear sign of addiction.
15. Elhai et al. (2017).
16. Jean Twenge (2017) shows survey data from large samples of US youth indicating a link between increased smartphone and social media use and increased rates of mental illness. It is worth pointing out (as Twenge does) that this is a correlation and so does not determine causation.

17. Whiting & Williams (2013). A paper from the 1980s suggests that people turned to television to alleviate boredom in much the same way we now turn to the Internet—essentially, to pass the time (Bryant & Zillmann, 1984).
18. Here we are using the term *simulacrum* as used by the French philosopher Jean Baudrillard (1994). He talks about a hyper-reality in which the thing being signified is supplanted by the signifier. If we were to simplify the idea a lot, we might say that the real world is replaced by a virtual, hyper-real world that feels more real than the real world. The Internet, social media, and the oxymoronically labeled "reality TV" all represent forms of simulacrum.
19. Thiele (1997). See also Aho (2007) for a similar analysis.
20. Thiele (1997), p. 505.
21. Cushman (1995).
22. Simmel (2012), p. 31.
23. Twenge (2017).
24. Arad et al. (2017); Tromholt (2016).
25. Yeykelis et al. (2014).
26. Damrad-Frye & Laird (1989).
27. Sexual boredom has been measured since the mid-1990s (Watt & Ewing, 1996). The work referred to here is from: Chaney & Chang (2005); Gana et al. (2000). One study suggests that men see sexual boredom as the price to pay for monogamy (Tunariu & Reavey, 2007).
28. See: https://www.nytimes.com/2018/01/17/world/europe/uk-britain-loneliness.html.
29. The Jo Cox Commission on Loneliness was established in honor of the British parliamentarian who was murdered during the Brexit campaign. Prior to her death she had worked to establish a commission to address loneliness in partnership with a range of NGOs. https://www.jocoxloneliness.org/.
30. Chin et al. (2017). Boredom and loneliness was touched on by many others before this study; Farmer & Sundberg (1986); Reissman et al. (1993); Spaeth et al. (2015).
31. Chan et al. (2018).

9. Just Go with the Flow

1. The movie *Free Solo* charts Honnold's free-solo climb of El Capitan, a 3,000-foot sheer granite rock face in Yosemite National Park.

2. Csikszentmihalyi (1975).
3. Csikszentmihalyi & Larson (2014). There are many ways to sample experiences, either in the short term (questions interrupting task performance) or long term (smartphone alerts asking what you are doing now and how you feel over a period of days, weeks, or months).
4. In fact, Czikszentmihalyi (1990) acknowledges that the term was derived from interviewees' descriptions of that state in which they felt "like floating" and being "carried on by the flow" (p. 40).
5. Marty-Dugas & Smilek (2019).
6. Struk et al. (2015).
7. Fahlman et al. (2013).
8. Pekrun et al. (2010, 2014). Chapter 8 established that increased novelty or complexity beyond a point will result in boredom. There is ample reason to believe that being overchallenged can lead to boredom. No doubt, in some situations, too much challenge may also lead to anxiety.
9. Blunt & Pychyl (2005); Ferrari (2000); Vodanovich & Rupp (1999).
10. Cheyne et al. (2006); Carriere et al. 2008; Hunter & Eastwood (2018, 2019); Malkovsky et al. (2012).
11. LeDoux & Pine (2016).
12. The full quotation from Alex Honnold can be found in Ferriss (2018): "I generally climb hard routes in the absence of fear. Though it's important to differentiate fear and risk. If there is a high level of risk, you should be feeling fear. It's a warning that there is real danger. Typically if I'm feeling a lot of fear, then I wait and prepare more, do whatever it takes to mitigate that, and then do the climb when I feel comfortable."
13. Fahlman et al. (2013); Gana et al. (2000); Harris (2000); Mercer-Lynn et al. (2013a); Seib & Vodanovich (1998).
14. For more of Honnold's climbing feats, see: http://www.alexhonnold.com/.
15. As quoted in Chancellor (2014).
16. Damrad-Frye & Laird (1989); Danckert & Allman (2005); Watt (1991).
17. Zakay (2014).
18. Renninger & Hidi (2015) note: "interest . . . refers to the psychological state of a person while engaging with some type of content . . . and *also* to the cognitive and affective motivational predisposition to re-engage with that content over time" (p. 8, emphasis added).
19. Dennett (2009).

20. Dennett calls this the 'joy of debugging'—using unexpected outcomes of jokes to fine-tune mental models of how the world *normally* functions. Hurley et al. (2011).
21. In addition to the calendar, more Dull Men can be found in the companion book. See Carlson (2015).
22. Gerstein (2003).
23. Some suggest we are hard-wired with a propensity to find certain things like complexity, ambiguity, and surprise interesting—but certainly not traffic cones (Berlyne, 1954, 1960, 1966, 1974; Hidi, 1990).
24. Nunoi and Yoshikawa (2016).
25. Zajonc (1968). Peretz et al. (1998) showed that even songs initially rated poorly received higher ratings when they were heard a second time.
26. Quoted in Schwarz (2018), p. 37; see also Zajonc (1968). Importantly, we suggest that when something becomes *too* familiar it runs the risk of becoming boring because it no longer engages our mental resources (e.g., Bornstein, 1989; Van den Bergh & Vrana, 1998).
27. It's likely not random what happens to be in front of us when the boredom push comes. In David's case, he was attending to traffic cones because of his work and an extrinsic motivation to resolve a patent dispute.
28. The quotation is widely attributed to Dorothy Parker.
29. Charnov (1976), Kurzban et al. (2013), Northcraft & Neale (1986).
30. In a recent computational approach to understanding boredom, curiosity, and exploration, researchers showed that an algorithm driven by boredom learned a novel environment more successfully than did an algorithm driven by curiosity. While the two states may both prompt exploration, it seems boredom may work more efficiently for learning (Yu et al., 2018).
31. We don't intend to use curiosity and interest interchangeably. It may be that curiosity spurs exploration that in turn leads to an activity that we consider interesting. Similarly, an activity we show interest in may spark our curiosity. It's also plausible that we have an interest—stamp collecting, let's say—that over time maintains our interest without *necessarily* sparking curiosity. Our point is that mental engagement comes in many forms, many of which don't require flow. On types of curiosity, see Berlyne (1954, 1966); Litman & Spielberger (2003); Reio et al. (2006). While terms differ among researchers, the general

dichotomy remains the same: informational versus sensory (Litman & Spielberger); epistemic versus perceptual (Berlyne).

32. Hunter et al. (2016); Kashdan et al. (2004); Pekrun et al. (2017); Reio et al. (2006).
33. Apathy is characterized by a total absence of desire, whereas relaxation is an *absence of unfulfilled desires.*
34. Sawin & Scerbo (1995) had high and low boredom-prone people do a boring attention task requiring them to detect infrequent flickers and told people to either "pay close attention" or "relax." Boredom ratings in the highly boredom prone came down to the same level as low boredom-prone people simply by telling them to relax!
35. Morita et al. (2007); Park et al. (2010).

Conclusion

1. The legend of the Gordian knot dates back to Ancient Greece and indicates an impossibly difficult problem with a simple solution, if only we could think outside the box. Legend has it that an oxcart of King Gordias was tied to a post with an impossibly complex series of knots. He who could untie the Gordian knot would become the next king. Apparently, Alexander first tried untying it before realizing he could break the knot by simply cutting it with his sword.
2. Fulfilling one's agency might not always result in desirable outcomes. Psychopaths may have a strong sense of agency, but few would condone their actions. Our claim is that agency-fulfilling actions are the best way to prevent boredom from becoming chronic and problematic. The question of how to behave in an ethical manner is another matter.
3. Russell (2012). "A generation that cannot endure boredom will be a generation of little men, of men unduly divorced from the slow process of nature, of men in whom every vital impulse slowly withers as though they were cut flowers in a vase" (p. 54).
4. Dixon et al. (2010, 2014). Mike Dixon and colleagues show that the "bells and whistles" associated with wins can be misleading. In multiline slot machine games if a gambler bets on seven lines and only wins on one of those lines, he loses money. Regardless, the machine still lights up and plays a happy song, fooling the hapless gambler into thinking he has won, keeping him engaged with the game. Dixon and

colleagues call this "losses disguised as wins," leading to "dark flow" (Dixon et al., 2018).

5. LePera (2011).
6. Koval & Todman (2015). People who practice mindfulness meditation regularly and report better mindfulness skills are better able to tolerate dull tasks without succumbing to boredom (Hallard, 2014; Petranker, 2018).
7. According to Buddhism, suffering is the result of our reaction to pain (of any sort, emotional or physical). This idea is captured in the maxim "Pain is inevitable, suffering is optional." This maxim applies to boredom as well as other aversive feelings.
8. The link mentioned in Chapter 5 between boredom and aggression is most prominently seen in a specific type of aggression—hostility (Isacescu et al., 2017). One way to interpret this is that when bored we exhibit hostility toward the world—the world is not enough, and we blame it for being so.
9. Warhol & Hackett (1988, p. 8).
10. Carl Honoré's book *In Praise of Slowness* (2004) has a similar message. Exploring the slow food movement and tantric sex, among other things, he doesn't suggest that *everything* should be done slowly. Rather, we should *choose* the pace of our lives.
11. Kierkegaard (1992, p. 214). "For the unhappy person is he who has his ideal, the content of this life, the fullness of his consciousness, his real nature in some way or other outside himself. The unhappy man is always absent from himself, never present to himself."
12. Brodsky (1995, pp. 109–110).
13. In his treatise on boredom, *The Pale King* (2011, p. 440), David Foster Wallace mused: "To be, in a word, unborable. . . . It is the key to modern life. If you are immune to boredom, there is literally nothing you cannot accomplish."
14. Brodsky (1995, pp. 110–111).
15. Nietzsche (2006, p. 385).

REFERENCES

Abramson, Edward E., and Shawn G. Stinson. 1977. "Boredom and eating in obese and non-obese individuals." *Addictive Behaviors* 2, no. 4: 181–185.

Aho, Kevin. 2007. "Simmel on acceleration, boredom, and extreme aesthesia." *Journal for the Theory of Social Behaviour* 37, no. 4: 447–462.

Arad, Ayala, Ohad Barzilay, and Maayan Perchick. 2017. "The impact of Facebook on social comparison and happiness: Evidence from a natural experiment." Unpublished manuscript, February 13. https://papers.ssrn.com/sol3/papers.cfm?abstract_id=2916158.

Arrigo, Bruce A., and Jennifer Leslie Bullock. 2008. "The psychological effects of solitary confinement on prisoners in supermax units: Reviewing what we know and recommending what should change." *International Journal of Offender Therapy and Comparative Criminology* 52, no. 6: 622–640.

Baddeley, Alan. 1996. "Exploring the central executive." *Quarterly Journal of Experimental Psychology Section A* 49, no. 1: 5–28.

Barbalet, Jack M. 1999. "Boredom and social meaning." *British Journal of Sociology* 50, no. 4: 631–646.

Bargdill, Richard. 2000. "The study of life boredom." *Journal of Phenomenological Psychology* 31, no. 2: 188–219.

Barmack, Joseph E. 1937. "Boredom and other factors in the physiology of mental effort: An exploratory study." *Archives of Psychology* 31: 1–83.

Barmack, Joseph E. 1938. "The effect of benzedrine sulfate (benzyl methyl carbinamine) upon the report of boredom and other factors." *Journal of Psychology* 5, no. 1: 125–133.

Barmack, Joseph E. 1939. "Studies on the psychophysiology of boredom: Part I. The effect of 15 mgs. of benzedrine sulfate and 60 mgs. of ephedrine hydrochloride on blood pressure, report of boredom and other factors." *Journal of Experimental Psychology* 25, no. 5: 494.

Barnett, Lynn A., and Sandra Wolf Klitzing. 2006. "Boredom in free time: Relationships with personality, affect, and motivation for different gender, racial and ethnic student groups." *Leisure Sciences* 28, no. 3: 223–244.

Bartone, Paul T., Amy B. Adler, and Mark A. Vaitkus. 1998. "Dimensions of psychological stress in peacekeeping operations." *Military Medicine* 163, no. 9: 587–593.

Baudrillard, Jean. 1994. *Simulacra and Simulation*. Ann Arbor: University of Michigan Press.

Bench, Shane W., and Heather C. Lench. 2013. "On the function of boredom." *Behavioral Sciences* 3, no. 3: 459–472.

Bengtsson, Tea Torbenfeldt. 2012. "Boredom and action: Experiences from youth confinement." *Journal of Contemporary Ethnography* 41, no. 5: 526–553.

Ben-Zeev, Dror, Michael A. Young, and Colin A. Depp. 2012. "Real-time predictors of suicidal ideation: Mobile assessment of hospitalized depressed patients." *Psychiatry Research* 197, no. 1–2: 55–59.

Berlyne, Daniel E. 1960. *Conflict, Arousal, and Curiosity*. New York: McGraw-Hill.

Berlyne, Daniel E. 1966. "Curiosity and exploration." *Science* 153, no. 3731: 25–33.

Berlyne, Daniel E. 1974. *Studies in the New Experimental Aesthetics: Steps toward an Objective Psychology of Aesthetic Appreciation*. Washington, DC: Hemisphere.

Berlyne, Daniel Ellis. 1954. "A theory of human curiosity." *British Journal of Psychology: General Section* 45, no. 3: 180–191.

Bernardi, Sylvia, and Stefano Pallanti. 2009. "Internet addiction: A descriptive clinical study focusing on comorbidities and dissociative symptoms." *Comprehensive Psychiatry* 50, no. 6: 510–516.

Bernstein, Haskell E. 1975. "Boredom and the ready-made life." *Social Research*: 512–537.

Best, John R., and Patricia H. Miller. 2010. "A developmental perspective on executive function." *Child Development* 81, no. 6: 1641–1660.

Biolcati, Roberta, Giacomo Mancini, and Elena Trombini. 2018. "Proneness to boredom and risk behaviors during adolescents' free time." *Psychological Reports* 121, no. 2: 303–323.

Bishop, Sheryl L. 2004. "Evaluating teams in extreme environments: From issues to answers." *Aviation, Space, and Environmental Medicine* 75, no. 7: C14–C21.

Blaszczynski, Alex, Neil McConaghy, and Anna Frankova. 1990. "Boredom proneness in pathological gambling." *Psychological Reports* 67, no. 1: 35–42.

Blunt, Allan, and Timothy A. Pychyl. 2005. "Project systems of procrastinators: A personal project-analytic and action control perspective." *Personality and Individual Differences* 38, no. 8: 1771–1780.

Blunt, Allan, and Timothy A. Pychyl. 1998. "Volitional action and inaction in the lives of undergraduate students: State orientation, procrastination and proneness to boredom." *Personality and Individual Differences* 24, no. 6: 837–846.

Boehm, Jim. 2006. *The Handbook for Exploding the Economic Myths of the Political Sound Bite.* West Conshohocken, PA: self published (Infinity Press).

Bolhuis, Jantina Elizabeth, Willem G. P. Schouten, Johan W. Schrama, and Victor M. Wiegant. 2006. "Effects of rearing and housing environment on behaviour and performance of pigs with different coping characteristics." *Applied Animal Behaviour Science* 101, no. 1–2: 68–85.

Bond, Frank W., Steven C. Hayes, Ruth A. Baer, Kenneth M. Carpenter, Nigel Guenole, Holly K. Orcutt, Tom Waltz, and Robert D. Zettle. 2011. "Preliminary psychometric properties of the Acceptance and Action Questionnaire–II: A revised measure of psychological inflexibility and experiential avoidance." *Behavior Therapy* 42, no. 4: 676–688.

Bonnaire, Céline, Michel Lejoyeux, and Roland Dardennes. 2004. "Sensation seeking in a French population of pathological gamblers: Comparison with regular and nongamblers." *Psychological Reports* 94, no. 3, suppl.: 1361–1371.

Bornmann, Lutz, and Rüdiger Mutz. 2015. "Growth rates of modern science: A bibliometric analysis based on the number of publications and cited references." *Journal of the Association for Information Science and Technology* 66, no. 11: 2215–2222.

Bornstein, Robert F. 1989. "Exposure and affect: Overview and meta-analysis of research, 1968–1987." *Psychological Bulletin* 106, no. 2: 265–289.

Boyle, Gregory J., Lisa M. Richards, and Anthony J. Baglioni Jr. 1993. "Children's Motivation Analysis Test (CMAT): An experimental manipulation of curiosity and boredom." *Personality and Individual Differences* 15, no. 6: 637–643.

Boys, Annabel, John Marsden, and John Strang. 2001. "Understanding reasons for drug use amongst young people: a functional perspective." *Health Education Research* 16, no. 4: 457–469.

Bradshaw, John W., Anne J. Pullen, and Nicola J. Rooney. 2015. "Why do adult dogs 'play'?" *Behavioural Processes* 110: 82–87.

Breazeal, Cynthia. 2009. "Role of expressive behaviour for robots that learn from people." *Philosophical Transactions of the Royal Society B: Biological Sciences* 364, no. 1535: 3527–3538.

Brissett, Dennis, and Robert P. Snow. 1993. "Boredom: Where the future isn't." *Symbolic Interaction* 16, no. 3: 237–256.

Britton, Annie, and Martin J. Shipley. 2010. "Bored to death?" *International Journal of Epidemiology* 39, no. 2: 370–371.

Brodsky, Joseph. 1995. "In praise of boredom." In *On Grief and Reason.* New York: Farrar Straus and Giroux.

Brown, Richard E. 2007. "Alfred McCoy, Hebb, the CIA and torture." *Journal of the History of the Behavioral Sciences* 43, no. 2: 205–213.

Bryant, Jennings, and Dolf Zillmann. 1984. "Using television to alleviate boredom and stress: Selective exposure as a function of induced excitational states." *Journal of Broadcasting and Electronic Media* 28, no. 1: 1–20.

Burda, Yuri, Harri Edwards, Deepak Pathak, Amos Storkey, Trevor Darrell, and Alexei A. Efros. 2018. "Large-scale study of curiosity-driven learning." Unpublished manuscript, August 13. https://arxiv.org/abs/1808.04355.

Burn, Charlotte C. 2017. "Bestial boredom: A biological perspective on animal boredom and suggestions for its scientific investigation." *Animal Behaviour* 130: 141–151.

Burney, C. 1952. *Solitary Confinement.* New York: Coward-McCann.

Burroughs, E. A. 1916. *The Fight for the Future.* London: Nisbet.

Byron, George Gordon, Lord. 2005. *Don Juan.* Rpt. ed., New York: Penguin Classics, 2005.

Cairns, Hugh, R. C. Oldfield, J. B. Pennybacker, and D. Whitteridge. 1941. "Akinetic mutism with an epidermoid cyst of the 3rd ventricle." *Brain* 64, no. 4: 273–290.

Caldwell, Linda L., Nancy Darling, Laura L. Payne, and Bonnie Dowdy. 1999. "'Why are you bored?': An examination of psychological and social control causes of boredom among adolescents." *Journal of Leisure Research* 31, no. 2: 103–121.

Caldwell, Linda L., and Edward A. Smith. 1995. "Health behaviors of leisure alienated youth." *Loisir et Société / Society and Leisure* 18, no. 1: 143–156.

Caldwell, L. L., and E. A. Smith. 2006. "Leisure as a context for youth development and delinquency prevention." *Australian and New Zealand Journal of Criminology* 39: 398–418.

Caldwell, Linda L., Edward A. Smith, and Ellen Weissinger. 1992. "Development of a leisure experience battery for adolescents: Parsimony, stability, and validity." *Journal of Leisure Research* 24, no. 4: 361–376.

Carlstead, Kathy. 1996. "Effects of captivity on the behavior of wild mammals." In *Wild Mammals in Captivity: Principles and Techniques,* ed. Devra Kleiman, Mary Allen, Susan Lumpkin, and Katerina Thompson. Chicago: University of Chicago Press.

Carlson, Leland. 2015. *Dull Men of Great Britain.* London: Ebury Press.

Carr, Nicholas. 2011. *The Shallows: What the Internet Is Doing to Our Brains.* New York: W. W. Norton.

Carriere, Jonathan S. A., J. Allan Cheyne, and Daniel Smilek. 2008. "Everyday attention lapses and memory failures: The affective consequences of mindlessness." *Consciousness and Cognition* 17, no. 3: 835–847.

Carroll, Douglas, and Justine A. A. Huxley. 1994. "Cognitive, dispositional, and psychophysiological correlates of dependent slot machine gambling in young people." *Journal of Applied Social Psychology* 24, no. 12: 1070–1083.

Chan, Christian S., Wijnand A. P. van Tilburg, Eric R. Igou, Cyanea Y. S. Poon, Katy Y. Y. Tam, Venus U. T. Wong, and S. K. Cheung. 2018. "Situational meaninglessness and state boredom: Cross-sectional and experience-sampling findings." *Motivation and Emotion* 42, no. 4: 555–565.

Chancellor, Will. 2014. "Alex Honnold." https://www.interviewmagazine.com/culture/alex-honnold.

Chaney, Michael P., and Catherine Y. Chang. 2005. "A trio of turmoil for Internet sexually addicted men who have sex with men: Boredom proneness, social connectedness, and dissociation." *Sexual Addiction and Compulsivity* 12, no. 1: 3–18.

Chapman, Alexander L., and Katherine L. Dixon-Gordon. 2007. "Emotional antecedents and consequences of deliberate self-harm and suicide attempts." *Suicide and Life-Threatening Behavior* 37, no. 5: 543–552.

Charnov, Eric L. 1976. "Optimal foraging, the marginal value theorem." *Theoretical Population Biology* 9, no. 2: 129–136.

Cheyne, James Allan, Jonathan S. A. Carriere, and Daniel Smilek. 2006. "Absent-mindedness: Lapses of conscious awareness and everyday cognitive failures." *Consciousness and Cognition* 15, no. 3: 578–592.

Chin, Alycia, Amanda Markey, Saurabh Bhargava, Karim S. Kassam, and George Loewenstein. 2017. "Bored in the USA: Experience sampling and boredom in everyday life." *Emotion* 17, no. 2: 359–368.

Ciocan, Cristian. 2010. "Heidegger and the problem of boredom." *Journal of the British Society for Phenomenology* 41, no. 1: 64–77.

Claes, Laurence, Walter Vandereycken, and Hans Vertommen. 2001. "Self-injurious behaviors in eating-disordered patients." *Eating Behaviors* 2, no. 3: 263–272.

Clarke, Dave, Samson Tse, Max W. Abbott, Sonia Townsend, Pefi Kingi, and Wiremu Manaia. 2007. "Reasons for starting and continuing gambling in a mixed ethnic community sample of pathological and non-problem gamblers." *International Gambling Studies* 7, no. 3: 299–313.

Coffman, Donna L., Linda L. Caldwell, and Edward A. Smith. 2012. "Introducing the at-risk average causal effect with application to HealthWise South Africa." *Prevention Science* 13, no. 4: 437–447.

Coman, Greg J., Graham D. Burrows, and Barry J. Evans. 1997. "Stress and anxiety as factors in the onset of problem gambling: Implications for treatment." *Stress Medicine* 13, no. 4: 235–244.

Conrad, Peter. 1997. "It's boring: Notes on the meanings of boredom in everyday life." *Qualitative Sociology* 20, no. 4: 465–475.

Conroy, Ronán M., Jeannette Golden, Isabelle Jeffares, Desmond O'Neill, and Hannah McGee. 2010. "Boredom-proneness, loneliness, social engagement and depression and their association with cognitive function in older people: A population study." *Psychology, Health and Medicine* 15, no. 4: 463–473.

Cook, Frederick Albert. 1909. *Through the First Antarctic Night, 1898–1899.* New York: Doubleday, Page.

Cotte, June, and Kathryn A. Latour. 2008. "Blackjack in the kitchen: Understanding online versus casino gambling." *Journal of Consumer Research* 35, no. 5: 742–758.

Coughlan, Gillian, Eric R. Igou, Wijnand A. P. van Tilburg, Elaine L. Kinsella, and Timothy D. Ritchie. 2019. "On boredom and perceptions

of heroes: A meaning-regulation approach to heroism." *Journal of Humanistic Psychology* 59, no. 4: 455–473.
Craig, Arthur D. 2009. "How do you feel—now? The anterior insula and human awareness." *Nature Reviews Neuroscience* 10, no. 1: 59–70.
Crockett, Amanda C., Samantha K. Myhre, and Paul D. Rokke. 2015. "Boredom proneness and emotion regulation predict emotional eating." *Journal of Health Psychology* 20, no. 5: 670–680.
Csikszentmihalyi, Mihaly, with contributions by I. Csikszentmihalyi. 1975. *Beyond Boredom and Anxiety.* San Francisco: Jossey-Bass.
Csikszentmihalyi, Mihaly. 1990. *Flow: The Psychology of Optimal Experience.* New York: Harper and Row.
Csikszentmihalyi, Mihaly, and Reed Larson. 2014. "Validity and reliability of the experience-sampling method." In *Flow and the Foundations of Positive Psychology,* 35–54. Dordrecht, NL: Springer.
Cushman, Philip. 1995. *Constructing the Self, Constructing America: A Cultural History of Psychotherapy.* Reading, MA: Addison-Wesley.
Dahl, Ronald E. 2004. "Adolescent brain development: A period of vulnerabilities and opportunities. Keynote address." *Annals of the New York Academy of Sciences* 1021, no. 1: 1–22.
Dahl, Ronald E. 2001. "Affect regulation, brain development, and behavioral / emotional health in adolescence." *CNS Spectrums* 6, no. 1: 60–72.
Dahlen, Eric R., Ryan C. Martin, Katie Ragan, and Myndi M. Kuhlman. 2004. "Boredom proneness in anger and aggression: Effects of impulsiveness and sensation seeking." *Personality and Individual Differences* 37, no. 8: 1615–1627.
Dal Mas, Dennis E., and Bianca C. Wittmann. 2017. "Avoiding boredom: Caudate and insula activity reflects boredom-elicited purchase bias." *Cortex* 92: 57–69.
Damrad-Frye, Robin, and James D. Laird. 1989. "The experience of boredom: The role of the self-perception of attention." *Journal of Personality and Social Psychology* 57, no. 2: 315.
Danckert, James A., and Ava-Ann A. Allman. 2005. "Time flies when you're having fun: Temporal estimation and the experience of boredom." *Brain and Cognition* 59, no. 3: 236–245.
Danckert, James, Tina Hammerschmidt, Jeremy Marty-Dugas, and Daniel Smilek. 2018a. "Boredom: Under-aroused and restless." *Consciousness and Cognition* 61: 24–37.

Danckert, James, and Julia Isacescu. 2017. "The bored brain: Insular cortex and the default mode network." Unpublished manuscript, September 27. https://psyarxiv.com/aqbcd/.

Danckert, James, and Colleen Merrifield. 2018. "Boredom, sustained attention and the default mode network." *Experimental Brain Research* 236, no. 9: 2507–2518.

Danckert, James, Jhotisha Mugon, Andriy Struk, and John D. Eastwood. 2018b. "Boredom: What is it good for?" In *The Function of Emotions,* ed. Heather C. Lench, 93–119. Cham: Springer.

Davies, A. Hudson. 1926. "Discussion on the physical and mental effects of monotony in modern industry." *British Medical Journal* 2, no. 3427: 472–479.

Dearborn, G. van N. 1932. "A case of congenital general pure analgesia." *Journal of Nervous and Mental Disease* 75: 612–615.

DeCarli, Charles, Joseph Massaro, Danielle Harvey, John Hald, Mats Tullberg, Rhoda Au, Alexa Beiser, Ralph D'Agostino, and Philip A. Wolf. 2005. "Measures of brain morphology and infarction in the Framingham heart study: Establishing what is normal." *Neurobiology of Aging* 26, no. 4: 491–510.

Deci, Edward L., and Richard M. Ryan. 1985. *Intrinsic Motivation and Self-Determination in Human Behavior.* New York: Plenum.

Deci, Edward L., and Richard M. Ryan. 2008. "Self-determination theory: A macrotheory of human motivation, development, and health." *Canadian Psychology / Psychologie canadienne* 49, no. 3: 182–185.

Dennett, Daniel. 2009. "Cute, Sexy, Sweet, Funny." Filmed March 15. TED video, 7:45. https://www.ted.com/talks/dan_dennett_cute_sexy_sweet_funny.

de Viggiani, Nick. 2007. "Unhealthy prisons: Exploring structural determinants of prison health." *Sociology of Health and Illness* 29, no. 1: 115–135.

Diamond, Adele. 2005. "Attention-deficit disorder (attention-deficit / hyperactivity disorder without hyperactivity): A neurobiologically and behaviorally distinct disorder from attention-deficit / hyperactivity disorder (with hyperactivity)." *Development and Psychopathology* 17, no. 3: 807–825.

Dickens, Charles. 2003. *Bleak House.* New York: Penguin Classics.

Dickens, Charles. 2005. *Bleak House.* Directed by Justin Chadwick and Susan White, screenplay by Andrew Davies, produced by Sally Haynes and Laura Mackie. BBC One, television serial.

Dickerson, Mark, John Hinchy, and John Fabre. 1987. "Chasing, arousal and sensation seeking in off-course gamblers." *British Journal of Addiction* 82, no. 6: 673–680.

Dixon, Mike J., Candice Graydon, Kevin A. Harrigan, Lisa Wojtowicz, Vivian Siu, and Jonathan A. Fugelsang. 2014. "The allure of multi-line games in modern slot machines." *Addiction* 109, no. 11: 1920–1928.

Dixon, Mike J., Kevin A. Harrigan, Rajwant Sandhu, Karen Collins, and Jonathan A. Fugelsang. 2010. "Losses disguised as wins in modern multi-line video slot machines." *Addiction* 105, no. 10: 1819–1824.

Dixon, Mike J., Madison Stange, Chanel J. Larche, Candice Graydon, Jonathan A. Fugelsang, and Kevin A. Harrigan. 2018. "Dark flow, depression and multiline slot machine play." *Journal of Gambling Studies* 34, no. 1: 73–84.

Drob, Sanford L., and Harold S. Bernard. 1988. "The bored patient: A developmental / existential perspective." *Psychotherapy Patient* 3, no. 3–4: 63–73.

Eakman, Aaron M. 2011. "Convergent validity of the Engagement in Meaningful Activities Survey in a college sample." *OTJR: Occupation, Participation and Health* 31, no. 1: 23–32.

Eastwood, John D., Carolina Cavaliere, Shelley A. Fahlman, and Adrienne E. Eastwood. 2007. "A desire for desires: Boredom and its relation to alexithymia." *Personality and Individual Differences* 42, no. 6: 1035–1045.

Eastwood, John D., Alexandra Frischen, Mark J. Fenske, and Daniel Smilek. 2012. "The unengaged mind: Defining boredom in terms of attention." *Perspectives on Psychological Science* 7, no. 5: 482–495.

Eastwood, John D., and Dana Gorelik. 2019. "Boredom is a feeling of thinking and a double-edged sword." In *Boredom Is in Your Mind*, ed. Josefa Ros Velasco, 55–70. Cham: Springer.

Eccleston, Chris, and Geert Crombez. 1999. "Pain demands attention: A cognitive–affective model of the interruptive function of pain." *Psychological Bulletin* 125, no. 3: 356–366.

Elhai, Jon D., Juanita K. Vasquez, Samuel D. Lustgarten, Jason C. Levine, and Brian J. Hall. 2018. "Proneness to boredom mediates relationships between problematic smartphone use with depression and anxiety severity." *Social Science Computer Review* 36, no. 6: 707–720.

Elpidorou, Andreas. 2014. "The bright side of boredom." *Frontiers in Psychology* 5: article 1245.

Elpidorou, Andreas. 2018. "The good of boredom." *Philosophical Psychology* 31, no. 3: 323–351.

Elpidorou, Andreas. 2017. "The moral dimensions of boredom: A call for research." *Review of General Psychology* 21, no. 1: 30–48.

Emerson, R. W. 1971. "Lecture on the Times," in *Collected Works of Ralph Waldo Emerson. Vol. 1: Nature, Addresses, and Lectures,* ed. R. E. Spiller and A. R. Ferguson. Cambridge, MA: Harvard University Press.

Fahlman, Shelley A., Kimberley B. Mercer-Lynn, David B. Flora, and John D. Eastwood. 2013. "Development and validation of the multidimensional state boredom scale." *Assessment* 20, no. 1: 68–85.

Fahlman, Shelley A., Kimberley B. Mercer, Peter Gaskovski, Adrienne E. Eastwood, and John D. Eastwood. 2009. "Does a lack of life meaning cause boredom? Results from psychometric, longitudinal, and experimental analyses." *Journal of Social and Clinical Psychology* 28, no. 3: 307–340.

Farmer, Richard, and Norman D. Sundberg. 1986. "Boredom proneness: The development and correlates of a new scale." *Journal of Personality Assessment* 50, no. 1: 4–17.

Farnworth, Louise. 1998. "Doing, being, and boredom." *Journal of Occupational Science* 5, no. 3: 140–146.

Favazza, Armando R. 1998. "The coming of age of self-mutilation." *Journal of Nervous and Mental Disease* 186, no. 5: 259–268.

Fenichel, Otto. 1953. *The Collected Papers of Otto Fenichel.* New York: W. W. Norton.

Fenichel, Otto. 1951. "On the psychology of boredom." *Organization and Pathology of Thought*: 349–361.

Ferrari, Joseph R. 2000. "Procrastination and attention: Factor analysis of attention deficit, boredomness, intelligence, self-esteem, and task delay frequencies." *Journal of Social Behavior and Personality* 15, no. 5, SPI: 185–196.

Ferrell, Jeff. 2004. "Boredom, crime and criminology." *Theoretical Criminology* 8, no. 3: 287–302.

Ferriss, Tim. 2018. "Alex Honnold talks to Tim Ferriss about fear and risk." https://www.outsideonline.com/2334851/alex-honnold-talks-tim-ferriss-about-fear-and-risk#close.

Fiske, Susan T., and Shelley E. Taylor. 1984. *Social Cognition: Topics in Social Psychology*. New York: Random House.

Fleming, Jennifer, Jennifer Sampson, Petrea Cornwell, Ben Turner, and Janell Griffin. 2012. "Brain injury rehabilitation: The lived experience of

inpatients and their family caregivers." *Scandinavian Journal of Occupational Therapy* 19, no. 2: 184–193.
Fogelman, Ken. 1976. "Bored eleven-year-olds." *British Journal of Social Work* 6, no. 2: 201–211.
Fox, Kieran C. R., Evan Thompson, Jessica R. Andrews-Hanna, and Kalina Christoff. 2014. "Is thinking really aversive? A commentary on Wilson et al.'s 'Just think: The challenges of the disengaged mind.'" *Frontiers in Psychology* 5: article 1427.
Frankl, Viktor. 1959. *Man's Search for Meaning.* Trans. Ilse Lasch. Boston: Beacon Press.
Frankl, Victor. 1978. *The Unheard Cry for Meaning: Psychotherapy and Humanism.* New York: Simon and Schuster.
Freeman, Frederick G., Peter J. Mikulka, Mark W. Scerbo, and Lorissa Scott. 2004. "An evaluation of an adaptive automation system using a cognitive vigilance task." *Biological Psychology* 67, no. 3: 283–297.
Frijda, Nico H. 2005. "Emotional experience." *Cognition and Emotion* 19: 473–497.
Frolova-Walker, Marina. 2004. "Stalin and the art of boredom." *Twentieth-century Music* 1, no. 1: 101–124.
Fromm, Erich. 1963. *The Dogma of Christ: And Other Essays on Religion, Psychology and Culture.* New York: Holt, Rinehart and Winston.
Fromm, Erich. 1955. *The Sane Society.* New York: Rinehart.
Galton, Francis. 1885. "The measure of fidget." *Nature* 32, no. 817: 174–175.
Gana, Kamel, Benedicte Deletang, and Laurence Metais. 2000. "Is boredom proneness associated with introspectiveness?" *Social Behavior and Personality: An International Journal* 28, no. 5: 499–504.
Gana, Kamel, Raphael Trouillet, Bettina Martin, and Leatitia Toffart. 2001. "The relationship between boredom proneness and solitary sexual behaviors in adults." *Social Behavior and Personality: An International Journal* 29, no. 4: 385–389.
Gasper, Karen, and Brianna L. Middlewood. 2014. "Approaching novel thoughts: Understanding why elation and boredom promote associative thought more than distress and relaxation." *Journal of Experimental Social Psychology* 52: 50–57.
Gerritsen, Cory J., Joel O. Goldberg, and John D. Eastwood. 2015. "Boredom proneness predicts quality of life in outpatients diagnosed with schizophrenia-spectrum disorders." *International Journal of Social Psychiatry* 61, no. 8: 781–787.

Gerritsen, Cory J., Maggie E. Toplak, Jessica Sciaraffa, and John Eastwood. 2014. "I can't get no satisfaction: Potential causes of boredom." *Consciousness and Cognition* 27: 27–41.

Gerstein, Mordicai. 2003. *The Man Who Walked between the Towers*. Brookfield, CT: Roaring Brook Press.

Giambra, Leonard M., Cameron J. Camp, and Alicia Grodsky. 1992. "Curiosity and stimulation seeking across the adult life span: Cross-sectional and 6-to 8-year longitudinal findings." *Psychology and Aging* 7, no. 1: 150–157.

Gill, Richard, Qixuan Chen, Debra D'Angelo, and Wendy K. Chung. 2014. "Eating in the absence of hunger but not loss of control behaviors are associated with 16p11. 2 deletions." *Obesity* 22, no. 12: 2625–2631.

Godin, Seth. 2007. *The Dip: A Little Book That Teaches You When to Quit (And When to Stick)*. New York: Penguin.

Goetz, Thomas, Anne C. Frenzel, Nathan C. Hall, Ulrike E. Nett, Reinhard Pekrun, and Anastasiya A. Lipnevich. 2014. "Types of boredom: An experience sampling approach." *Motivation and Emotion* 38, no. 3: 401–419.

Goldberg, Yael K. and James Danckert. 2013. "Traumatic brain injury, boredom and depression." *Behavioral Sciences* 3, no. 3: 434–444.

Goldberg, Yael K., John D. Eastwood, Jennifer LaGuardia, and James Danckert. "Boredom: An emotional experience distinct from apathy, anhedonia, or depression." 2011. *Journal of Social and Clinical Psychology* 30, no. 6: 647–666.

Gosselin, Frédéric, and Philippe G. Schyns. 2003. "Superstitious perceptions reveal properties of internal representations." *Psychological Science* 14, no. 5: 505–509.

Greenberg, Jeff, Sander Leon Koole, and Thomas A. Pyszczynski, eds. 2004. *Handbook of Experimental Existential Psychology*. New York: Guilford Press.

Greenson, Ralph R. 1953. "On boredom." *Journal of the American Psychoanalytic Association* 1, no. 1: 7–21.

Gross, Dominik, and Gereon Schäfer. 2011. "Egas Moniz (1874–1955) and the 'invention' of modern psychosurgery: A historical and ethical reanalysis under special consideration of Portuguese original sources." *Neurosurgical Focus* 30, no. 2: E8.

Hadfield, Chris. 2013. *An Astronaut's Guide to Life on Earth*. New York: Little, Brown.

Hafner, Katie, and Matthew Lyon. 1998. *Where wizards stay up late: The origins of the Internet*. Simon and Schuster, 1998.

Hallard, Robert Ian. 2014. "Mindfulness meditation practice can make concentration feel a little easier." *Cumbria Partnership Journal of Research Practice and Learning* 4: 17–22.

Haller, Max, Markus Hadler, and Gerd Kaup. 2013. "Leisure time in modern societies: A new source of boredom and stress?" *Social Indicators Research* 111, no. 2: 403–434.

Hamilton, Jean A. 1981. "Attention, personality, and the self-regulation of mood: Absorbing interest and boredom." *Progress in Experimental Personality Research* 10, no. 28: 281–315.

Hamilton, Jean A., Richard J. Haier, and Monte S. Buchsbaum. 1984. "Intrinsic enjoyment and boredom coping scales: Validation with personality, evoked potential and attention measures." *Personality and Individual Differences* 5, no. 2: 183–193.

Harden, K. Paige, and Elliot M. Tucker-Drob. 2011. "Individual differences in the development of sensation seeking and impulsivity during adolescence: Further evidence for a dual systems model." *Developmental Psychology* 47, no. 3: 739–746.

Harris, Mary B. 2000. "Correlates and characteristics of boredom proneness and boredom 1." *Journal of Applied Social Psychology* 30, no. 3: 576–598.

Havermans, Remco C., Linda Vancleef, Antonis Kalamatianos, and Chantal Nederkoorn. 2015. "Eating and inflicting pain out of boredom." *Appetite* 85: 52–57.

Hayes, Steven C., Kirk Strosahl, Kelly G. Wilson, Richard T. Bissett, Jacqueline Pistorello, Dosheen Toarmino, Melissa A. Polusny, et al. 2004. "Measuring experiential avoidance: A preliminary test of a working model." *Psychological Record* 54, no. 4: 553–578.

Healy, Sean Desmond. 1984. *Boredom, Self, and Culture*. Rutherford, NJ: Fairleigh Dickinson University Press.

Hebb, Donald O. 1980. "Donald O. Hebb." In *A History of Psychology in Autobiography*, vol. 7, ed. Gardner Lindzey, 273–303 (San Francisco: W. H. Freeman).

Heinrich, Joseph, Steven J. Heine, and Ara Norenzayan. 2010. "The weirdest people in the world." *Behavioral and Brain Sciences* 33, no. 2–3: 61–83.

Heron, Woodburn. 1957. "The pathology of boredom." *Scientific American* 196: 52–57.

Hesse, Hermann. 1951. *Siddhartha*. Trans. Hilda Rosner. New York: New Directions.

Hidi, Suzanne. 1990. "Interest and its contribution as a mental resource for learning." *Review of Educational Research* 60, no. 4: 549–571.

Hing, Nerilee, and Helen Breen. 2001. "Profiling lady luck: An empirical study of gambling and problem gambling amongst female club members." *Journal of Gambling Studies* 17, no. 1: 47–69.

Hitchcock, Edward M., William N. Dember, Joel S. Warm, Brian W. Moroney, and Judi E. See. 1999. "Effects of cueing and knowledge of results on workload and boredom in sustained attention." *Human Factors* 41, no. 3: 365–372.

Homer. 1962 / 1990. *The Odyssey*. Trans. Robert Fitzgerald. New York: Knopf Doubleday.

Honoré, Carl. 2004. *In Praise of Slowness: How a Worldwide Movement Is Challenging the Cult of Speed*. San Francisco: HarperSanFrancisco.

Hopley, Anthony A. B., Kevin Dempsey, and Richard Nicki. 2012. "Texas Hold'em online poker: A further examination." *International Journal of Mental Health and Addiction* 10, no. 4: 563–572.

Hopley, Anthony A. B., and Richard M. Nicki. 2010. "Predictive factors of excessive online poker playing." *Cyberpsychology, Behavior, and Social Networking* 13, no. 4: 379–385.

Hunt, Laurence T., and Benjamin Y. Hayden. 2017. "A distributed, hierarchical and recurrent framework for reward-based choice." *Nature Reviews Neuroscience* 18, no. 3: 172–182.

Hunter, Andrew G., and John D. Eastwood. 2018. "Does state boredom cause failures of attention? Examining the relations between trait boredom, state boredom, and sustained attention." *Experimental Brain Research* 236, no. 9: 2483–2492.

Hunter, Andrew G., and John D. Eastwood. 2019. "Idle hands, listless minds: Unpacking the dynamics of boredom and attention." Paper presented at the 29th meeting of the Canadian Society for Brain, Behaviour and Cognitive Science, Waterloo, ON, June.

Hunter, Jennifer A., E. H. Abraham, A. G. Hunter, L. C. Goldberg, and J. D. Eastwood. 2016. "Personality and Boredom Proneness in the Prediction of Creativity and Curiosity." *Thinking Skills and Creativity* 22: 48–57.

Hurley, Matthew M., Daniel C. Dennett, and Reginald B. Adams Jr. 2011. *Inside Jokes: Using Humor to Reverse-Engineer the Mind*. Cambridge, MA: MIT Press.

Ice, Gillian Harper. 2002. "Daily life in a nursing home: Has it changed in 25 years?" *Journal of Aging Studies* 16, no. 4: 345–359.

Inman, Alice, Kenneth L. Kirsh, and Steven D. Passik. 2003. "A pilot study to examine the relationship between boredom and spirituality in cancer patients." *Palliative and Supportive Care* 1, no. 2: 143–151.

Inzlicht, Michael, and Lisa Legault. 2014. "No pain, no gain: How distress underlies effective self-control (and unites diverse social psychological phenomena)." In *Motivation and Its Regulation: The Control Within,* ed. Joseph P. Forgas and Eddie Harmon-Jones, 115–132. New York: Psychology Press.

Isacescu, Julia, and James Danckert. 2018. "Exploring the relationship between boredom proneness and self-control in traumatic brain injury (TBI)." *Experimental Brain Research* 236, no. 9: 2493–2505.

Isacescu, Julia, Andriy A. Struk, and James Danckert. 2017. "Cognitive and affective predictors of boredom proneness." *Cognition and Emotion* 31, no. 8: 1741–1748.

Iso-Ahola, Seppo E., and Edward D. Crowley. 1991. "Adolescent substance abuse and leisure boredom." *Journal of Leisure Research* 23, no. 3: 260–271.

James, William. 1900. *On Some of Life's Ideals.* New York: H. Holt.

Jiang, Yang, Joann Lianekhammy, Adam Lawson, Chunyan Guo, Donald Lynam, Jane E. Joseph, Brian T. Gold, and Thomas H. Kelly. 2009. "Brain responses to repeated visual experience among low and high sensation seekers: Role of boredom susceptibility." *Psychiatry Research: Neuroimaging* 173, no. 2: 100–106.

Joireman, Jeff, Jonathan Anderson, and Alan Strathman. 2003. "The aggression paradox: Understanding links among aggression, sensation seeking, and the consideration of future consequences." *Journal of Personality and Social Psychology* 84, no. 6: 1287–1302.

Kahneman, Daniel. 1973. *Attention and Effort.* Englewood Cliffs, NJ: Prentice-Hall.

Kangas, David. 2008. "Kierkegaard." In *The Oxford Handbook of Religion and Emotion,* ed. John Corrigan. New York: Oxford University Press.

Kashdan, Todd B., Paul Rose, and Frank D. Fincham. 2004. "Curiosity and exploration: Facilitating positive subjective experiences and personal growth opportunities." *Journal of Personality Assessment* 82, no. 3: 291–305.

Kass, Steven J., and Stephen J. Vodanovich. 1990. "Boredom proneness: Its relationship to Type A behavior pattern and sensation seeking." *Psychology: A Journal of Human Behavior*: 7–16.

Kass, Steven J., Stephen J. Vodanovich, Claudia J. Stanny, and Tiffany M. Taylor. 2001. "Watching the clock: Boredom and vigilance performance." *Perceptual and Motor Skills* 92, no. 3 suppl.: 969–976.

Kass, Steven J., J. Craig Wallace, and Stephen J. Vodanovich. 2003. "Boredom proneness and sleep disorders as predictors of adult attention deficit scores." *Journal of Attention Disorders* 7, no. 2: 83–91.

Kenah, Katrina, Julie Bernhardt, Toby Cumming, Neil Spratt, Julie Luker, and Heidi Janssen. 2018. "Boredom in patients with acquired brain injuries during inpatient rehabilitation: A scoping review." *Disability and Rehabilitation* 40, no. 22: 2713–2722.

Kierkegaard, Søren. 1992. *Either / Or: A Fragment of Life*. Ed. Victor Eremita, abridged and trans. Alastair Hannay. London: Penguin.

Klapp, Orrin Edgar. 1986. *Overload and Boredom: Essays on the Quality of Life in the Information Society*. New York: Greenwood Press.

Koball, Afton M., Molly R. Meers, Amy Storfer-Isser, Sarah E. Domoff, and Dara R. Musher-Eizenman. 2012. "Eating when bored: Revision of the Emotional Eating Scale with a focus on boredom." *Health Psychology* 31, no. 4: 521.

Korzenny, Felipe, and Kimberly Neuendorf. 1980. "Television viewing and self-concept of the elderly." *Journal of Communication* 30, no. 1: 71–80.

Koval, Samuel R., and McWelling Todman. 2015. "Induced boredom constrains mindfulness: An online demonstration." *Psychology and Cognitive Science—Open Journal* 1, no. 1: 1–9.

Kracauer, Siegfried. 1995. *The Mass Ornament: Weimar Essays*. Ed. and trans. Thomas Y. Levin. Cambridge, MA: Harvard University Press.

Kreutzer, Jeffrey S., Ronald T. Seel, and Eugene Gourley. 2001. "The prevalence and symptom rates of depression after traumatic brain injury: a comprehensive examination." *Brain Injury* 15, no. 7: 563–576.

Krotava, Iryna, and McWelling Todman. 2014. "Boredom severity, depression and alcohol consumption in Belarus." *Journal of Psychology and Behavioral Science* 2, no. 1: 73–83.

Kruglanski, Arie W., Erik P. Thompson, E. Tory Higgins, Nadir Atash, Antonia Pierro, James Y. Shah, and Scott Spiegel. 2000. "To 'do the right thing' or to 'just do it': Locomotion and assessment as distinct self-regulatory imperatives." *Journal of Personality and Social Psychology* 79: 793–815.

Kuhl, Julius. 1994. "Action versus state orientation: Psychometric properties of the Action Control Scale (ACS-90)." *Volition and Personality: Action versus State Orientation* 47: 47–59.

Kuhl, Julius. 1981. "Motivational and functional helplessness: The moderating effect of state versus action orientation." *Journal of Personality and Social Psychology* 40, no. 1: 155–170.

Kuhl, Julius. 1985. "Volitional mediators of cognition-behavior consistency: Self-regulatory processes and action versus state orientation." In *Action Control, from Cognition to Behavior*, ed. Julius Kuhl and Jürgen Beckmann, 101–128. Berlin: Springer.

Kuhn, Reinhard Clifford. 1976. *The Demon of Noontide: Ennui in Western Literature*. Princeton: Princeton University Press.

Kunzendorf, Robert G., and Franz Buker. 2008. "Does existential meaning require hope, or is interest enough?" *Imagination, Cognition and Personality* 27, no. 3: 233–243.

Kurzban, Robert, Angela Duckworth, Joseph W. Kable, and Justus Myers. 2013. "An opportunity cost model of subjective effort and task performance." *Behavioral and Brain Sciences* 36, no. 6: 661–679.

Kustermans, Jorg, and Erik Ringmar. 2011. "Modernity, boredom, and war: a suggestive essay." *Review of International Studies* 37, no. 4: 1775–1792.

Landon, P. Bruce, and Peter Suedfeld. 1969. "Information and meaningfulness needs in sensory deprivation." *Psychonomic Science* 17, no. 4: 248.

Larson, Reed W. 1990. "Emotions and the creative process; anxiety, boredom, and enjoyment as predictors of creative writing." *Imagination, Cognition and Personality* 9, no. 4: 275–292.

Larson, Reed W., and Maryse H. Richards. 1991. "Boredom in the middle school years: Blaming schools versus blaming students." *American Journal of Education* 99, no. 4: 418–443.

Lebedev, Valentin Vital'evich. 1988. *Diary of a Cosmonaut: 211 Days in Space*. Trans. Luba Diangar, ed. Daniel Puckett and C. W. Harrison. College Station, TX: PhytoResource Research, Inc., Information Service.

LeDoux, Joseph E., and Daniel S. Pine. 2016. "Using neuroscience to help understand fear and anxiety: a two-system framework." *American Journal of Psychiatry* 173: 1083–1093.

Lee, Christine M., Clayton Neighbors, and Briana A. Woods. 2007. "Marijuana motives: Young adults' reasons for using marijuana." *Addictive Behaviors* 32, no. 7: 1384–1394.

Lehr, Evangeline, and McWelling Todman. 2009. "Boredom and boredom proneness in children: Implications for academic and social adjustment." *Self-Regulation and Social Competence: Psychological Studies in Identity, Achievement and Work-Family Dynamics,* ed. M. Todman, 75–90. Athens: ATNIER Press.

Leon, Gloria R., and Karen Chamberlain. 1973. "Emotional arousal, eating patterns, and body image as differential factors associated with varying success in maintaining a weight loss." *Journal of Consulting and Clinical Psychology* 40, no. 3: 474–480.

Leong, Frederick T. L., and Gregory R. Schneller. 1993. "Boredom proneness: Temperamental and cognitive components." *Personality and Individual Differences* 14, no. 1: 233–239.

LePera, Nicole. 2011. "Relationships between boredom proneness, mindfulness, anxiety, depression, and substance use." *New School Psychology Bulletin* 8, no. 2: 15–25.

Lewinsky, Hilde. 1943. "Boredom." *British Journal of Educational Psychology* 13, no. 3: 147–152.

Lewis, Kerrie P. 2000. "A comparative study of primate play behaviour: Implications for the study of cognition." *Folia Primatologica* 71, no. 6: 417–421.

Lipps, Theodor. 1906. *Leitfaden der Psychologie*. Leipzig: Wilhelm Engelmann.

Litman, Jordan A., and Charles D. Spielberger. 2003. "Measuring epistemic curiosity and its diversive and specific components." *Journal of Personality Assessment* 80, no. 1: 75–86.

Lowenstein, Otto, and Irene E. Loewenfeld. 1952. "Disintegration of central autonomic regulation during fatigue and its reintegration by psychosensory controlling mechanisms. I. Disintegration. Pupillographic studies." *Journal of Nervous and Mental Disease* 115: 121–145.

Lowenstein, Otto, and Irene E. Loewenfeld. 1951. "Types of central autonomic innervation and fatigue: Pupillographic studies." *AMA Archives of Neurology and Psychiatry* 66, no. 5: 580–599.

Luria, Aleksandr R. 1973. *The Working Brain: An Introduction to Neuropsychology*. Trans. Basil Haigh. New York: Basic Books.

MacDonald, Douglas A., and Daniel Holland. 2002. "Spirituality and boredom proneness." *Personality and Individual Differences* 32, no. 6: 1113–1119.

Mackworth, Norman H. 1948. "The breakdown of vigilance during prolonged visual search." *Quarterly Journal of Experimental Psychology* 1, no. 1: 6–21.

Maddi, Salvatore R. 1967. "The existential neurosis." *Journal of Abnormal Psychology* 72, no. 4: 311–325.

Maddi, Salvatore R. 1970. "The search for meaning." In *Nebraska Symposium on Motivation,* vol. 17, 134–183. Lincoln: University of Nebraska Press.

Malkovsky, Ela, Colleen Merrifield, Yael Goldberg, and James Danckert. 2012. "Exploring the relationship between boredom and sustained attention." *Experimental Brain Research* 221, no. 1: 59–67.

Mann, Sandi, and Rebekah Cadman. 2014. "Does being bored make us more creative?" *Creativity Research Journal* 26, no. 2: 165–173.

Marin, Robert S., and Patricia A. Wilkosz. 2005. "Disorders of diminished motivation." *Journal of Head Trauma Rehabilitation* 20, no. 4: 377–388.

Martin, Marion, Gaynor Sadlo, and Graham Stew. 2006. "The phenomenon of boredom." *Qualitative Research in Psychology* 3, no. 3: 193–211.

Marty-Dugas, Jeremy, and Daniel Smilek. 2019. "Deep, effortless concentration: Re-examining the flow concept and exploring relations with inattention, absorption, and personality." *Psychological Research* 83, no. 8: 1760–1777.

Mathiak, Krystyna Anna, Martin Klasen, Mikhail Zvyagintsev, René Weber, and Klaus Mathiak. 2013. "Neural networks underlying affective states in a multimodal virtual environment: contributions to boredom." *Frontiers in Human Neuroscience* 7: article 820.

Matthies, Swantje, Alexandra Philipsen, and Jennifer Svaldi. 2012. "Risky decision making in adults with ADHD." *Journal of Behavior Therapy and Experimental Psychiatry* 43, no. 3: 938–946.

McDonald, William. 2009. "Kierkegaard's Demonic Boredom." In *Essays on Boredom and Modernity,* ed. Barbara Dalle Pezze and Carlo Salzani, 61–85. Leiden, NL: Brill.

McIntosh, James, Fiona MacDonald, and Neil McKeganey. 2005. "The reasons why children in their pre and early teenage years do or do not use illegal drugs." *International Journal of Drug Policy* 16, no. 4: 254–261.

McIvor, Arthur J. 1987a. "Employers, the government, and industrial fatigue in Britain, 1890–1918." *Occupational and Environmental Medicine* 44, no. 11: 724–732.

McIvor, Arthur J. 1987b. "Manual work, technology, and industrial health, 1918–39." *Medical History* 31, no. 2: 160–189.

McLeod, Carol R., and Stephen J. Vodanovich. 1991. "The relationship between self-actualization and boredom proneness." *Journal of Social Behavior and Personality* 6, no. 5: 137–146.

McNeilly, Dennis P., and William J. Burke. 2000. "Late life gambling: The attitudes and behaviors of older adults." *Journal of Gambling Studies* 16, no. 4: 393–415.

Meagher, Rebecca K., Dana L. M. Campbell, and Georgia J. Mason. 2017. "Boredom-like states in mink and their behavioural correlates: A replicate study." *Applied Animal Behaviour Science* 197: 112–119.

Meagher, Rebecca K., and Georgia J. Mason. 2012. "Environmental enrichment reduces signs of boredom in caged mink." *PLoS One* 7, no. 11: e49180.

Medaglia, John Dominic, Fabio Pasqualetti, Roy H. Hamilton, Sharon L. Thompson-Schill, and Danielle S. Bassett. 2017. "Brain and cognitive reserve: Translation via network control theory." *Neuroscience and Biobehavioral Reviews* 75: 53–64.

Mega, Michael S., and Robert C. Cohenour. 1997. "Akinetic mutism: Disconnection of frontal-subcortical circuits." *Neuropsychiatry, Neuropsychology, and Behavioral Neurology* 10: 254–259.

Melton, Amanda M. A., and Stefan E. Schulenberg. 2009. "A confirmatory factor analysis of the boredom proneness scale." *Journal of Psychology* 143, no. 5: 493–508.

Melton, Amanda M. A., and Stefan E. Schulenberg. 2007. "On the relationship between meaning in life and boredom proneness: Examining a logotherapy postulate." *Psychological Reports* 101, no. 3, suppl.: 1016–1022.

Mercer, Kimberley B., and John D. Eastwood. 2010. "Is boredom associated with problem gambling behaviour? It depends on what you mean by 'boredom.'" *International Gambling Studies* 10, no. 1: 91–104.

Mercer-Lynn, Kimberley B., Rachel J. Bar, and John D. Eastwood. 2014. "Causes of boredom: The person, the situation, or both?" *Personality and Individual Differences* 56: 122–126.

Mercer-Lynn, Kimberley B., David B. Flora, Shelley A. Fahlman, and John D. Eastwood. 2013a. "The measurement of boredom: Differences between existing self-report scales." *Assessment* 20, no. 5: 585–596.

Mercer-Lynn, Kimberley B., Jennifer A. Hunter, and John D. Eastwood. 2013b. "Is trait boredom redundant?" *Journal of Social and Clinical Psychology* 32, no. 8: 897–916.

Merrifield, Colleen, and James Danckert. 2014. "Characterizing the psychophysiological signature of boredom." *Experimental Brain Research* 232, no. 2: 481–491.

Miller, Jacqueline A., Linda L. Caldwell, Elizabeth H. Weybright, Edward A. Smith, Tania Vergnani, and Lisa Wegner. 2014. "Was Bob Seger right? Relation between boredom in leisure and [risky] sex." *Leisure Sciences* 36, no. 1: 52–67.

Miyake, Akira, Naomi P. Friedman, Michael J. Emerson, Alexander H. Witzki, Amy Howerter, and Tor D. Wager. 2000. "The unity and diversity of executive functions and their contributions to complex "frontal lobe" tasks: A latent variable analysis." *Cognitive Psychology* 41, no. 1: 49–100.

Morita, Emi, S. Fukuda, Jun Nagano, N. Hamajima, H. Yamamoto, Y. Iwai, T. Nakashima, H. Ohira, and T. J. P. H. Shirakawa. 2007. "Psychological effects of forest environments on healthy adults: Shinrin-yoku (forest-air bathing, walking) as a possible method of stress reduction." *Public Health* 121, no. 1: 54–63.

Moynihan, Andrew B., Eric R. Igou, and Wijnand A. P. van Tilburg. 2017. "Boredom increases impulsiveness." *Social Psychology* 48, no. 5: 293–309.

Moynihan, Andrew B., Wijnand A. P. van Tilburg, Eric R. Igou, Arnaud Wisman, Alan E. Donnelly, and Jessie B. Mulcaire. 2015. "Eaten up by boredom: Consuming food to escape awareness of the bored self." *Frontiers in Psychology* 6: article 369.

Mugon, Jhotisha, Andriy Struk, and James Danckert. 2018. "A failure to launch: Regulatory modes and boredom proneness." *Frontiers in Psychology* 9: article 1126.

Münsterberg, Hugo. 1913. *Psychology and Industrial Efficiency.* Boston: Mifflin.

"National Survey of American Attitudes on Substance Abuse VIII: Teens and Parents." 2003. National Center on Addiction and Substance Abuse, Columbia University, August.

Nault, Jean-Charles. 2015. *The Noonday Devil: Acedia, the Unnamed Evil of Our Times.* San Francisco: Ignatius Press.

Nederkoorn, Chantal, Linda Vancleef, Alexandra Wilkenhöner, Laurence Claes, and Remco C. Havermans. 2016. "Self-inflicted pain out of boredom." *Psychiatry Research* 237: 127–132.

Németh, G. 1988. "Some theoretical and practical aspects of the disturbances of consciousness with special reference to akinetic mutism." *Functional Neurology* 3, no. 1: 9–28.

Nett, Ulrike E., Thomas Goetz, and Lia M. Daniels. 2010. "What to do when feeling bored? Students' strategies for coping with boredom." *Learning and Individual Differences* 20, no. 6: 626–638.

Nett, Ulrike E., Thomas Goetz, and Nathan C. Hall. 2011. "Coping with boredom in school: An experience sampling perspective." *Contemporary Educational Psychology* 36, no. 1: 49–59.

Newell, Susan E., Priscilla Harries, and Susan Ayers. 2012. "Boredom proneness in a psychiatric inpatient population." *International Journal of Social Psychiatry* 58, no. 5: 488–495.

Ng, Andy H., Yong Liu, Jian-zhi Chen, and John D. Eastwood. 2015. "Culture and state boredom: A comparison between European Canadians and Chinese." *Personality and Individual Differences* 75: 13–18.

Nichols, Laura A., and Richard Nicki. 2004. "Development of a psychometrically sound internet addiction scale: A preliminary step." *Psychology of Addictive Behaviors* 18, no. 4: 381.

Nietzsche, Friedrich. 2006. *Human, All-Too-Human.* Trans. Helen Zimmern and Paul V. Cohn. Mineola, NY: Dover.

Northcraft, Gregory B., and Margaret A. Neale. 1986. "Opportunity costs and the framing of resource allocation decisions." *Organizational Behavior and Human Decision Processes* 37, no. 3: 348–356.

Nower, Lia, and Alex Blaszczynski. 2006. "Impulsivity and pathological gambling: A descriptive model." *International Gambling Studies* 6, no. 1: 61–75.

Nunoi, Masato, and Sakiko Yoshikawa. 2016. "Deep processing makes stimuli more preferable over long durations." *Journal of Cognitive Psychology* 28, no. 6: 756–763.

Oddy, Michael, Michael Humphrey, and David Uttley. 1978. "Subjective impairment and social recovery after closed head injury." *Journal of Neurology, Neurosurgery and Psychiatry* 41, no. 7: 611–616.

O'Hanlon, James F. 1981. "Boredom: Practical consequences and a theory." *Acta Psychologica* 49, no. 1: 53–82.

Orcutt, James D. 1984. "Contrasting effects of two kinds of boredom on alcohol use." *Journal of Drug Issues* 14, no. 1: 161–173.

Palinkas, Lawrence A. 2003. "The psychology of isolated and confined environments: Understanding human behavior in Antarctica." *American Psychologist* 58, no. 5: 353–363.

Palinkas, Lawrence A., Eric Gunderson, Albert W. Holland, Christopher Miller, and Jeffrey C. Johnson. 2000. "Predictors of behavior and performance in extreme environments: The Antarctic space analogue program." *Aviation, Space, and Environmental Medicine* 71: 619–625.

Paliwoda, Daniel. 2010. *Melville and the Theme of Boredom*. Jefferson, NC: McFarland.

Park, Bum Jin, Yuko Tsunetsugu, Tamami Kasetani, Takahide Kagawa, and Yoshifumi Miyazaki. 2010. "The physiological effects of Shinrin-yoku (taking in the forest atmosphere or forest bathing): Evidence from field experiments in 24 forests across Japan." *Environmental Health and Preventive Medicine* 15, no. 1: 18–26.

Passik, Steven D., Alice Inman, Kenneth Kirsh, Dale Theobald, and Pamela Dickerson. 2003. "Initial validation of a scale to measure purposelessness, understimulation, and boredom in cancer patients: Toward a redefinition of depression in advanced disease." *Palliative and Supportive Care* 1, no. 1: 41–50.

Patterson, Ian, Shane Pegg, and Roberta Dobson-Patterson. 2000. "Exploring the links between leisure boredom and alcohol use among youth in rural and urban areas of Australia." *Journal of Park and Recreation Administration* 18, no. 3: 53–75.

Pattyn, Nathalie, Xavier Neyt, David Henderickx, and Eric Soetens. 2008. "Psychophysiological investigation of vigilance decrement: boredom or cognitive fatigue?" *Physiology and Behavior* 93, no. 1–2: 369–378.

Pekrun, Reinhard, Thomas Goetz, Lia M. Daniels, Robert H. Stupnisky, and Raymond P. Perry. 2010. "Boredom in achievement settings: Exploring control-value antecedents and performance outcomes of a neglected emotion." *Journal of Educational Psychology* 102, no. 3: 531–549.

Pekrun, Reinhard, Nathan C. Hall, Thomas Goetz, and Raymond P. Perry. 2014. "Boredom and academic achievement: Testing a model of reciprocal causation." *Journal of Educational Psychology* 106, no. 3: 696–710.

Pekrun, Reinhard, Elisabeth Vogl, Krista R. Muis, and Gale M. Sinatra. 2017. "Measuring emotions during epistemic activities: The Epistemically-Related Emotion Scales." *Cognition and Emotion* 31, no. 6: 1268–1276.

Pellis, Sergio, and Vivien Pellis. 2009. *The Playful Brain: Venturing to the Limits of Neuroscience*. Oxford: Oneworld.

Peretz, Isabelle, Lise Gagnon, and Bernard Bouchard. 1998. "Music and emotion: Perceptual determinants, immediacy, and isolation after brain damage." *Cognition* 68, no. 2: 111–141.

Petranker, Rotem. 2018. "Sitting with it: Examining the relationship between mindfulness, sustained attention, and boredom." M.A. thesis, York University.

Pettiford, Jasmine, Rachel V. Kozink, Avery M. Lutz, Scott H. Kollins, Jed E. Rose, and F. Joseph McClernon. 2007. "Increases in impulsivity following smoking abstinence are related to baseline nicotine intake and boredom susceptibility." *Addictive Behaviors* 32, no. 10: 2351–2357.

Phillips, Adam. 1994. *On Kissing, Tickling, and Being Bored: Psychoanalytic Essays on the Unexamined Life*. Cambridge, MA: Harvard University Press.

Piaget, Jean. 1999. *Judgment and Reasoning in the Child.* International Library of Psychology, Book 23, rpt. ed. New York: Routledge.

Piko, Bettina F., Thomas A. Wills, and Carmella Walker. 2007. "Motives for smoking and drinking: Country and gender differences in samples of Hungarian and US high school students." *Addictive Behaviors* 32, no. 10: 2087–2098.

Pitrat, Jacques. 2009. *Artificial Beings: The Conscience of a Conscious Machine.* Hoboken, NJ: John Wiley.

Potegal, Michael, and Dorothy Einon. 1989. "Aggressive behaviors in adult rats deprived of playfighting experience as juveniles." *Developmental Psychobiology* 22, no. 2: 159–172.

Pribram, Karl H., and Diane McGuinness. 1975. "Arousal, activation, and effort in the control of attention." *Psychological Review* 82, no. 2: 116–149.

Protheroe, S. M. 1991. "Congenital insensitivity to pain." *Journal of the Royal Society of Medicine* 84, no. 9: 558–559.

Quay, Herbert C. 1965. "Psychopathic personality as pathological stimulation-seeking." *American Journal of Psychiatry* 122, no. 2: 180–183.

Raffaelli, Quentin, Caitlin Mills, and Kalina Christoff. 2018. "The knowns and unknowns of boredom: A review of the literature." *Experimental Brain Research* 236, no. 9: 2451–2462.

Raposa, Michael L. 1999. *Boredom and the Religious Imagination*. Charlottesville: University Press of Virginia.

Raz, Mical. 2013. "Alone again: John Zubek and the troubled history of sensory deprivation research." *Journal of the History of the Behavioral Sciences* 49, no. 4: 379–395.

Reio, Thomas G., Jr., Joseph M. Petrosko, Albert K. Wiswell, and Juthamas Thongsukmag. 2006. "The measurement and conceptualization of curiosity." *Journal of Genetic Psychology* 167, no. 2: 117–135.

Reissman, Charlotte, Arthur Aron, and Merlynn R. Bergen. 1993. "Shared activities and marital satisfaction: Causal direction and self-expansion versus boredom." *Journal of Social and Personal Relationships* 10, no. 2: 243–254.

Renninger, K. Ann, and Suzanne Hidi. 2015. *The Power of Interest for Motivation and Engagement*. New York: Routledge.

Riem, Madelon M. E., Alexandra Voorthuis, Marian J. Bakermans-Kranenburg, and Marinus H. van Ijzendoorn. 2014. "Pity or peanuts? Oxytocin induces different neural responses to the same infant crying labeled as sick or bored." *Developmental Science* 17, no. 2: 248–256.

Risko, Evan F., and Sam J. Gilbert. 2016. "Cognitive offloading." *Trends in Cognitive Sciences* 20, no. 9: 676–688.

Rizvi, Sakina J., Diego A. Pizzagalli, Beth A. Sproule, and Sidney H. Kennedy. 2016. "Assessing anhedonia in depression: Potentials and pitfalls." *Neuroscience and Biobehavioral Reviews* 65: 21–35.

Romand, David. 2015. "Theodor Waitz's theory of feelings and the rise of affective sciences in the mid-19th century." *History of Psychology* 18, no. 4: 385–400.

Rupp, Deborah E., and Stephen J. Vodanovich. 1997. "The role of boredom proneness in self-reported anger and aggression." *Journal of Social Behavior and Personality* 12, no. 4: 925–936.

Russell, Bertrand. 2012. *The Conquest of Happiness*. Abingdon, UK: Routledge.

Russo, Mary F., Benjamin B. Lahey, Mary Anne G. Christ, Paul J. Frick, Keith McBurnett, Jason L. Walker, Rolf Loeber, Magda Stouthamer-Loeber, and Stephanie Green. 1991. "Preliminary development of a sensation seeking scale for children." *Personality and Individual Differences* 12, no. 5: 399–405.

Russo, Mary F., Garnett S. Stokes, Benjamin B. Lahey, Mary Anne G. Christ, Keith McBurnett, Rolf Loeber, Magda Stouthamer-Loeber, and Stephanie M. Green. 1993. "A sensation seeking scale for children: Further refinement and psychometric development." *Journal of Psychopathology and Behavioral Assessment* 15, no. 2: 69–86.

Ryan, Richard M., and Edward L. Deci. 2000. "Self-determination theory and the facilitation of intrinsic motivation, social development, and well-being." *American Psychologist* 55, no. 1: 68–78.

Sandal, Gro M., G. R. Leon, and Lawrence Palinkas. 2006. "Human challenges in polar and space environments." In *Life in Extreme*

Environments, ed. R. Amils, C. Ellis-Evans, and H. G. Hinghofer-Szalkay, 399–414. Dordrecht, NL: Springer.

Sansone, Carol, Charlene Weir, Lora Harpster, and Carolyn Morgan. 1992. "Once a boring task always a boring task? Interest as a self-regulatory mechanism." *Journal of Personality and Social Psychology* 63, no. 3: 379–390.

Sawin, David A., and Mark W. Scerbo. 1995. "Effects of instruction type and boredom proneness in vigilance: Implications for boredom and workload." *Human Factors* 37, no. 4: 752–765.

Scerbo, Mark W. 1998. "What's so boring about vigilance?" In *Viewing Psychology as a Whole: The Integrative Science of William N. Dember,* ed. R. R. Hoffman, M. F. Sherrick, and J. S. Warm, 145–166. Washington, DC: American Psychological Association.

Scherer, Klaus R. 1997. "College life on-line: Healthy and unhealthy Internet use." *Journal of College Student Development* 38: 655–665.

Scherer, Klaus R. 2005. "What are emotions? And how can they be measured?" *Social Science Information* 44, no. 4: 695–729.

Schopenhauer, Arthur. 1995. *The World as Will and Idea,* ed. David Berman, trans. J. Berman. London: Everyman.

Schwarz, Norbert. 2018. "Of fluency, beauty, and truth." In *Metacognitive Diversity: An Interdisciplinary Approach,* ed. Joelle Proust and Martin Fortier, 25–46. Oxford: Oxford University Press.

Scuteri, Angelo, Luigi Palmieri, Cinzia Lo Noce, and Simona Giampaoli. 2005. "Age-related changes in cognitive domains: A population-based study." *Aging Clinical and Experimental Research* 17, no. 5: 367–373.

Seel, Ronald T., and Jeffrey S. Kreutzer. 2003. "Depression assessment after traumatic brain injury: An empirically based classification method." *Archives of Physical Medicine and Rehabilitation* 84, no. 11: 1621–1628.

Seib, Hope M., and Stephen J. Vodanovich. 1998. "Cognitive correlates of boredom proneness: The role of private self-consciousness and absorption." *Journal of Psychology* 132, no. 6: 642–652.

Sharp, Erin Hiley, and Linda L. Caldwell. 2005. "Understanding adolescent boredom in leisure: A longitudinal analysis of the roles of parents and motivation." In *Eleventh Canadian Congress on Leisure Research, Malaspina University College Nanaimo, British Columbia.*

Sharp, Erin Hiley, Donna L. Coffman, Linda L. Caldwell, Edward A. Smith, Lisa Wegner, Tania Vergnani, and Catherine Mathews. 2011. "Predicting substance use behavior among South African adolescents:

The role of leisure experiences across time." *International Journal of Behavioral Development* 35, no. 4: 343–351.

Sharpe, Lynda. 2011. "So you think you know why animals play. . . ." *Scientific American* guest blog, May 17.

Shiota, Michelle N., Dacher Keltner, and Amanda Mossman. 2007. "The nature of awe: Elicitors, appraisals, and effects on self-concept." *Cognition and Emotion* 21, no. 5: 944–963.

Shoalts, Adam. 2017. *A History of Canada in Ten Maps.* Toronto: Allen Lane,.

Shuman-Paretsky, Melissa, Vance Zemon, Frederick W. Foley, and Roee Holtzer. 2017. "Development and validation of the State-Trait Inventory of Cognitive Fatigue in community-dwelling older adults." *Archives of Physical Medicine and Rehabilitation* 98, no. 4: 766–773.

Simmel, Georg. 2012. "The Metropolis and Mental Life." In *The Urban Sociology Reader,* ed. Jan Lin and Christopher Mele, 23–31. New York: Routledge.

Sirigu, Angela, and Jean-René Duhamel. 2016. "Reward and decision processes in the brains of humans and nonhuman primates." *Dialogues in Clinical Neuroscience* 18, no. 1: 45–53.

Smith, Adam. 1976. *An Inquiry into the Nature and Causes of the Wealth of Nations,* 2 vols. Ed. Edwin Cannan. London, 1776; Chicago: University of Chicago Press.

Smith, Edward A., and Linda L. Caldwell. 1989. "The perceived quality of leisure experiences among smoking and nonsmoking adolescents." *Journal of Early Adolescence* 9, no. 1–2: 153–162.

Smith, Peter Scharff. 2006. "The effects of solitary confinement on prison inmates: A brief history and review of the literature." *Crime and Justice* 34, no. 1: 441–528.

Smith, Richard P. 1981. "Boredom: A review." *Human Factors* 23, no. 3: 329–340.

Solomon, Andrew. 2001. *The Noonday Demon: An Atlas of Depression.* New York: Scribner.

Spacks, Patricia Meyer. 1995. *Boredom: The Literary History of a State of Mind.* Chicago: University of Chicago Press.

Spaeth, Michael, Karina Weichold, and Rainer K. Silbereisen. 2015. "The development of leisure boredom in early adolescence: Predictors and longitudinal associations with delinquency and depression." *Developmental Psychology* 51, no. 10: 1380–1394.

Stanovich, Keith. 2011. *Rationality and the Reflective Mind.* New York: Oxford University Press.

Steele, Rachel, Paul Henderson, Frances Lennon, and Donna Swinden. 2013. "Boredom among psychiatric in-patients: Does it matter?" *Advances in Psychiatric Treatment* 19, no. 4: 259–267.

Steinberg, Laurence. 2005. "Cognitive and affective development in adolescence." *Trends in Cognitive Sciences* 9, no. 2: 69–74.

Stevenson, M. F. 1983. "The captive environment: Its effect on exploratory and related behavioural responses in wild animals." In *Exploration in Animals and Humans,* ed. John Archer and Lynda I. A. Birke, 176–197. Wokingham, UK: Van Nostrand Reinhold.

Stickney, Marcella I., Raymond G. Miltenberger, and Gretchen Wolff. 1999. "A descriptive analysis of factors contributing to binge eating." *Journal of Behavior Therapy and Experimental Psychiatry* 30, no. 3: 177–189.

Struk, Andriy, A. Scholer, and J. Danckert. 2015. "Perceived control predicts engagement and diminished boredom." Presentation at the Canadian Society for Brain, Behaviour and Cognitive Science.

Struk, Andriy A., Jonathan S. A. Carriere, J. Allan Cheyne, and James Danckert. 2017. "A short Boredom Proneness Scale: Development and psychometric properties." *Assessment* 24, no. 3: 346–359.

Struk, Andriy A., Abigail A. Scholer, and James Danckert. 2016. "A self-regulatory approach to understanding boredom proneness." *Cognition and Emotion* 30, no. 8: 1388–1401.

Sulea, Coralia, Ilona Van Beek, Paul Sarbescu, Delia Virga, and Wilmar B. Schaufeli. 2015. "Engagement, boredom, and burnout among students: Basic need satisfaction matters more than personality traits." *Learning and Individual Differences* 42: 132–138.

Svendsen, Lars. 2005. *A Philosophy of Boredom.* London: Reaktion Books.

Tabatabaie, Ashkan Fakhr, Mohammad Reza Azadehfar, Negin Mirian, Maryam Noroozian, Ahmad Yoonessi, Mohammad Reza Saebipour, and Ali Yoonessi. 2014. "Neural correlates of boredom in music perception." *Basic and Clinical Neuroscience* 5, no. 4: 259–266.

Taylor, Christopher A., Jeneita M. Bell, Matthew J. Breiding, and Likang Xu. "Traumatic brain injury–related emergency department visits, hospitalizations, and deaths—United States, 2007 and 2013." *MMWR Surveillance Summaries* 66, no. 9 (2017): 1–16. DOI: http://dx.doi.org/10.15585/mmwr.ss6609a1.

Teo, Thomas. 2007. "Local institutionalization, discontinuity, and German textbooks of psychology, 1816–1854." *Journal of the History of the Behavioral Sciences* 43, no. 2: 135–157.

Thackray, Richard I., J. Powell Bailey, and R. Mark Touchstone. 1977. "Physiological, subjective, and performance correlates of reported boredom and monotony while performing a simulated radar control task." In *Vigilance: Theory, Operational Performance, and Physiological Correlates,* ed. Robert R. Mackie, 203–215. Boston: Springer.

Theobold, Dale E., Kenneth L. Kirsh, Elizabeth Holtsclaw, Kathleen Donaghy, and Steven D. Passik. 2003. "An open label pilot study of citalopram for depression and boredom in ambulatory cancer patients." *Palliative and Supportive Care* 1, no. 1: 71–77.

Thiele, Leslie Paul. 1997. "Postmodernity and the routinization of novelty: Heidegger on boredom and technology." *Polity* 29, no. 4: 489–517.

Todman, McWelling. 2003. "Boredom and psychotic disorders: Cognitive and motivational issues." *Psychiatry: Interpersonal and Biological Processes* 66, no. 2: 146–167.

Todman, McWelling. 2013. "The dimensions of state boredom: Frequency, duration, unpleasantness, consequences and causal attributions." *Dimensions* 1: 32–40.

Todman, McWelling, Daniel Sheypuk, Kristin Nelson, Jason Evans, Roger Goldberg, and Evangeline Lehr. 2008. "Boredom, hallucination-proneness and hypohedonia in schizophrenia and schizoaffective disorder." In *Schizoaffective Disorders: International Perspectives on Understanding, Intervention and Rehabilitation,* ed. Kam-Shing Yip. Hauppauge, NY: Nova Science Publishers.

Tolinski, Brad, and Alan Di Perna. 2016. *Play It Loud: An Epic History of the Style, Sound, and Revolution of the Electric Guitar*. New York: Anchor Doubleday.

Tolor, Alexander, and Marlene C. Siegel. 1989. "Boredom proneness and political activism." *Psychological Reports* 65, no. 1: 235–240.

Tolstoy, Leo. 1899. *Anna Karénina*. Trans. Nathan Haskell Dole. New York: Thomas Y. Crowell.

Toohey, Peter. 2011. *Boredom: A Lively History*. New Haven, CT: Yale University Press.

Trevorrow, Karen, and Susan Moore. 1998. "The association between loneliness, social isolation and women's electronic gaming machine gambling." *Journal of Gambling Studies* 14, no. 3: 263–284.

Tromholt, Morten. 2016. "The Facebook experiment: Quitting Facebook leads to higher levels of well-being." *Cyberpsychology, Behavior, and Social Networking* 19, no. 11: 661–666.

Tunariu, Aneta D., and Paula Reavey. 2007. "Common patterns of sense making: A discursive reading of quantitative and interpretative data on sexual boredom." *British Journal of Social Psychology* 46, no. 4: 815–837.

Turing, Alan M. 1950. "Computing machinery and intelligence." *Mind: A Quarterly Review of Psychology and Philosophy* 59: 433–460.

Turner, Nigel E., Masood Zangeneh, and Nina Littman-Sharp. 2006. "The experience of gambling and its role in problem gambling." *International Gambling Studies* 6, no. 2: 237–266.

Twenge, Jean M. 2017. *iGen: Why Today's Super-Connected Kids Are Growing Up Less Rebellious, More Tolerant, Less Happy—and Completely Unprepared for Adulthood—and What That Means for the Rest of Us*. New York: Simon and Schuster.

Tze, Virginia M. C., Robert M. Klassen, and Lia M. Daniels. 2014. "Patterns of boredom and its relationship with perceived autonomy support and engagement." *Contemporary Educational Psychology* 39, no. 3: 175–187.

Uddin, Lucina Q. 2015. "Salience processing and insular cortical function and dysfunction." *Nature Reviews Neuroscience* 16, no. 1: 55–61.

Ulrich, Martin, Johannes Keller, and Georg Grön. 2015. "Neural signatures of experimentally induced flow experiences identified in a typical fMRI block design with BOLD imaging." *Social Cognitive and Affective Neuroscience* 11, no. 3: 496–507.

Ulrich, Martin, Johannes Keller, Klaus Hoenig, Christiane Waller, and Georg Grön. 2014. "Neural correlates of experimentally induced flow experiences." *Neuroimage* 86 (2014): 194–202.

Valenzuela, Michael J., and Perminder Sachdev. 2006. "Brain reserve and cognitive decline: A non-parametric systematic review." *Psychological Medicine* 36, no. 8: 1065–1073.

Van den Bergh, Omer, and Scott R. Vrana. 1998. "Repetition and boredom in a perceptual fluency/attributional model of affective judgements." *Cognition and Emotion* 12, no. 4: 533–553.

van Tilburg, Wijnand A. P., and Eric R. Igou. 2017. "Boredom begs to differ: Differentiation from other negative emotions." *Emotion* 17, no. 2: 309–322.

van Tilburg, Wijnand A. P., and Eric R. Igou. 2016. "Going to political extremes in response to boredom." *European Journal of Social Psychology* 46, no. 6: 687–699.
van Tilburg, Wijnand A. P., and Eric R. Igou. 2012. "On boredom: Lack of challenge and meaning as distinct boredom experiences." *Motivation and Emotion* 36, no. 2: 181–194.
van Tilburg, Wijnand A. P., and Eric R. Igou. 2011. "On boredom and social identity: A pragmatic meaning-regulation approach." *Personality and Social Psychology Bulletin* 37, no. 12: 1679–1691.
van Tilburg, Wijnand A. P., Eric R. Igou, and Constantine Sedikides. 2013. "In search of meaningfulness: Nostalgia as an antidote to boredom." *Emotion* 13, no. 3: 450–461.
Vodanovich, Stephen J. 2003. "Psychometric measures of boredom: A review of the literature." *Journal of Psychology* 137, no. 6: 569–595.
Vodanovich, Stephen J., and Deborah E. Rupp. 1999. "Are procrastinators prone to boredom?" *Social Behavior and Personality* 27, no. 1: 11–16.
Vodanovich, Stephen J., Kathryn M. Verner, and Thomas V. Gilbride. 1991. "Boredom proneness: Its relationship to positive and negative affect." *Psychological Reports* 69, no. 3, suppl.: 1139–1146.
Vodanovich, Stephen J., J. Craig Wallace, and Steven J. Kass. 2005. "A confirmatory approach to the factor structure of the Boredom Proneness Scale: Evidence for a two-factor short form." *Journal of Personality Assessment* 85, no. 3: 295–303.
Vodanovich, Stephen J., and John D. Watt. 2016. "Self-report measures of boredom: An updated review of the literature." *Journal of Psychology* 150, no. 2: 196–228.
Waitz, Theodore. 1849. *Lehrbuch der Psychologie als Naturwissenschaft* [Textbook of psychology as a natural science]. Braunschweig, Germany: Vieweg.
Walfish, Steven, and Tuesdai A. Brown. 2009. "Self-assessed emotional factors contributing to increased weight in presurgical male bariatric patients." *Bariatric Nursing and Surgical Patient Care* 4, no. 1: 49–52.
Wallace, David Foster. 2011. *The Pale King: An Unfinished Novel.* New York: Little, Brown.
Wallace, J. Craig, Steven J. Kass, and Claudia J. Stanny. 2002. "The cognitive failures questionnaire revisited: Dimensions and correlates." *Journal of General Psychology* 129, no. 3: 238–256.
Wallace, J. Craig, Stephen J. Vodanovich, and Becca M. Restino. 2003. "Predicting cognitive failures from boredom proneness and daytime

sleepiness scores: An investigation within military and undergraduate samples." *Personality and Individual Differences* 34, no. 4: 635–644.

Wangh, Martin. 1975. "Boredom in psychoanalytic perspective." *Social Research* 42: 538–550.

Wangh, Martin. 1979. "Some psychoanalytic observations on boredom." *International Journal of Psycho-Analysis* 60: 515–526.

Wardley, Kenneth Jason. 2012. "'A weariness of the flesh': Towards a theology of boredom and fatigue." In *Intensities: Philosophy, Religion and the Affirmation of Life,* ed. Katharine Sarah Moody and Steven Shakespeare, 117–136. Burlington, VT: Ashgate.

Warhol, Andy, and Pat Hackett. 1988. *Andy Warhol's Party Book*. New York: Crown.

Warriner, Amy Beth, and Karin R. Humphreys. 2008. "Learning to fail: Reoccurring tip-of-the-tongue states." *Quarterly Journal of Experimental Psychology* 61, no. 4: 535–542.

Watt, John D. 1991. "Effect of boredom proneness on time perception." *Psychological Reports* 69, no. 1: 323–327.

Watt, John D., and Jackie E. Ewing. 1996. "Toward the development and validation of a measure of sexual boredom." *Journal of Sex Research* 33, no. 1: 57–66.

Watt, John D., and Stephen J. Vodanovich. 1992. "Relationship between boredom proneness and impulsivity." *Psychological Reports* 70, no. 3: 688–690.

Wegner, Lisa. 2011. "Through the lens of a peer: Understanding leisure boredom and risk behaviour in adolescence." *South African Journal of Occupational Therapy* 41, no. 1: 19–23.

Wegner, Lisa, and Alan J. Flisher. 2009. "Leisure boredom and adolescent risk behaviour: A systematic literature review." *Journal of Child and Adolescent Mental Health* 21, no. 1: 1–28.

Weinberg, Warren A., and Roger A. Brumback. 1990. "Primary disorder of vigilance: A novel explanation of inattentiveness, daydreaming, boredom, restlessness, and sleepiness." *Journal of Pediatrics* 116, no. 5: 720–725.

Weinstein, Lawrence, Xiaolin Xie, and Charalambos C. Cleanthous. 1995. "Purpose in life, boredom, and volunteerism in a group of retirees." *Psychological Reports* 76, no. 2: 482.

Weissinger, Ellen. 1995. "Effects of boredom on self-reported health." *Loisir et societé / Society and Leisure* 18, no. 1: 21–32.

Weissinger, Ellen, Linda L. Caldwell, and Deborah L. Bandalos. 1992. "Relation between intrinsic motivation and boredom in leisure time." *Leisure Sciences* 14, no. 4: 317–325.

Wemelsfelder, Françoise. 1985. "Animal boredom: Is a scientific study of the subjective experiences of animals possible?" In *Advances in Animal Welfare Science 1984/85*, ed. M. W. Fox and L. D. Mickley, 115–154. Dordrecht: Martinus Nijhoff/Kluwer.

Wemelsfelder, Françoise. 2005. "Animal boredom: Understanding the tedium of confined lives." In *Mental Health and Well-Being in Animals,* ed. F. D. McMillan, 77–91. Oxford: Blackwell.

Wemelsfelder, Françoise. 1990. "Boredom and laboratory animal welfare." In *The Experimental Animal and Biomedical Research,* ed. Bernard Rollin, 243–272. Boca Raton: CRC Press.

Wemelsfelder, Françoise. 1993. "The concept of animal boredom and its relationship to stereotyped behaviour." In *Stereotypic Behavior: Fundamentals and Applications to Animal Welfare,* ed. A. B. Lawrence and J. Rushen, 95–96. Tucson, AZ: CAB International.

Westgate, Erin C., and Timothy D. Wilson. 2018 "Boring thoughts and bored minds: The MAC model of boredom and cognitive engagement." *Psychological Review* 125, no. 5: 689–713.

Weybright, Elizabeth H., Linda L. Caldwell, Nilam Ram, Edward A. Smith, and Lisa Wegner. 2015. "Boredom prone or nothing to do? Distinguishing between state and trait leisure boredom and its association with substance use in South African adolescents." *Leisure Sciences* 37, no. 4: 311–331.

White, A. 1998. "Ho hum: A phenomenology of boredom." *Journal of the Society for Existential Analysis* 9: 69–81.

White, Robert W. 1959. "Motivation reconsidered: The concept of competence." *Psychological Review* 66, no. 5: 297–333.

Whiting, Anita, and David Williams. 2013. "Why people use social media: A uses and gratifications approach." *Qualitative Market Research: An International Journal* 16, no. 4: 362–369.

Willging, Cathleen E., Gilbert A. Quintero, and Elizabeth A. Lilliott. 2014. "Hitting the wall: Youth perspectives on boredom, trouble, and drug use dynamics in rural New Mexico." *Youth and Society* 46, no. 1: 3–29.

Williams, D. J., and Mary Liz Hinton. 2006. "Leisure experience, prison culture, or victimization? Sex offenders report on prison gambling." *Victims and Offenders* 1, no. 2: 175–192.

Wilson, Timothy D., David A. Reinhard, Erin C. Westgate, Daniel T. Gilbert, Nicole Ellerbeck, Cheryl Hahn, Casey L. Brown, and Adi Shaked. 2014. "Just think: The challenges of the disengaged mind." *Science* 345, no. 6192: 75–77.

Wink, Paul, and Karen Donahue. 1995. "Implications of college-age narcissism for psychosocial functioning at midlife: Findings from a longitudinal study of women." *Journal of Adult Development* 2, no. 2: 73–85.

Wink, Paul, and Karen Donahue. 1997. "The relation between two types of narcissism and boredom." *Journal of Research in Personality* 31, no. 1: 136–140.

Winokur, Jon, ed. 2005. *Ennui to Go: The Art of Boredom*. Seattle: Sasquatch Books.

Witte, Kim, and William A. Donohue. 2000. "Preventing vehicle crashes with trains at grade crossings: The risk seeker challenge." *Accident Analysis and Prevention* 32, no. 1: 127–139.

Wood, Richard T. A., Mark D. Griffiths, and Jonathan Parke. 2007. "Acquisition, development, and maintenance of online poker playing in a student sample." *Cyberpsychology and Behavior* 10, no. 3: 354–361.

Wyatt, Stanley, and James A. Fraser. 1929. "The effects of monotony in work—a preliminary enquiry." Oxford: H.M. Stationery Office.

Wyatt, Stanley, and James N. Langdon. 1937. *Fatigue and Boredom in Repetitive Work*. London: H.M. Stationery Office.

Yeykelis, Leo, James J. Cummings, and Byron Reeves. 2014. "Multitasking on a single device: Arousal and the frequency, anticipation, and prediction of switching between media content on a computer." *Journal of Communication* 64, no. 1: 167–192.

Young, Kimberly S. 1998. "Internet addiction: The emergence of a new clinical disorder." *Cyberpsychology and Behavior* 1, no. 3: 237–244.

Young, Kimberly S., and Robert C. Rogers. 1998. "The relationship between depression and Internet addiction." *Cyberpsychology and Behavior* 1, no. 1: 25–28.

Yu, Yen, Acer Y. C. Chang, and Ryota Kanai. 2018. "Boredom-driven curious learning by homeo-heterostatic value gradients." *Frontiers in Neurorobotics* 12.

Yunis, Harvey, ed. 2011. *Plato, Phaedrus.* Cambridge Greek and Latin Classics. New York: Cambridge University Press.

Zajonc, Robert B. 1968. "Attitudinal effects of mere exposure." *Journal of Personality and Social Psychology* 9, no. 2: 1–27.

Zakay, Dan. 2014. "Psychological time as information: The case of boredom." *Frontiers in Psychology* 5: article 917.

Ziervogel, C. F., Najma Ahmed, A. J. Flisher, and B. A. Robertson. 1997. "Alcohol misuse in South African male adolescents: A qualitative investigation." *International Quarterly of Community Health Education* 17, no. 1: 25–41.

Zondag, Hessel J. 2013. "Narcissism and boredom revisited: An exploration of correlates of overt and covert narcissism among Dutch university students." *Psychological Reports* 112, no. 2: 563–576.

Zuckerman, Marvin. 1993. "P-impulsive sensation seeking and its behavioral, psychophysiological and biochemical correlates." *Neuropsychobiology* 28, no. 1–2: 30–36.

Zuckerman, Marvin. 1979. *Sensation Seeking: Beyond the Optimal Level of Arousal.* Hillsdale, NJ: Lawrence Erlbaum Associates.

ACKNOWLEDGMENTS

JD

I would like to thank all of my graduate students who have conducted the boredom research in the lab over the past decade—Yael Goldberg, Julia Isacescu, Colleen Merrifield, Jhotisha Mugon, and Andriy Struk. More than just foot soldiers in this endeavor, you have all shaped my thinking in important ways. There has been an army of undergraduates who also contributed, but I'll reserve special mention for Ava-Ann Allman—our paper in 2005 kicked it off for the lab. I would like to thank my colleagues, who also shape the work and should be credited for the best ideas and immune to the most egregious errors—Abby Scholer, Ian McGregor, and Dan Smilek. A special thanks to Colin Ellard for advice and insight on this whole book writing thingy. I would like to thank our editor Janice Audet and everyone at Harvard University Press for guiding us so expertly through this process. Finally, I would like to thank my family for supporting many late nights stuck at my computer, for putting up with me when bored, and for generally helping keep that to a minimum. As always, Stacey can't be thanked enough for holding it all together.

JDE

Thanks to the keen graduate students and postdoctoral fellows who have explored boredom with me—Veerpal Bambrah, Carol Cavaliere, Shelley Fahlman, Alexandra Frischen, Cory Gerritsen, Dana Gorelik, Andrew Hunter, Jennifer Hunter, Chia-Fen Hsu, Sanaz Mehranvar, Kimberley Mercer-Lynn, Andy Ng, and Rotem Petranker. In particular, Shelley, who set the foundation for the boredom lab, and Alexandra, who boosted our momentum, deserve special acknowledgment. Special thanks also goes to

Jennifer and Andrew, who provided feedback on early drafts of some of this book. Colleagues Mark Fenske, Peter Gaskovski, Ian McGregor, Ian Newby-Clark, and Dan Smilek have played a formative role in my professional life. Dan and Mark significantly shaped my thinking about boredom in particular and must be recognized for their immeasurable contribution, yet not faulted for any of my own missteps. Special thanks also go to Maggie Toplak, who provided invaluable moral support as we both wrestled with our respective writing projects over the last couple of years. Collaboration and dialogue are central to what keeps me engaged in the research enterprise, and I am very thankful to each and every one of you. Janice Audet, our editor, and everyone else at Harvard University Press provided just the right amount of freedom and direction to keep us on track. I am thankful for their guidance during the writing process. Finally, and most importantly, I thank Adrienne for indulging and supporting me as I pursued this project. Words can't express how grateful I am to have you by my side.

ILLUSTRATION CREDITS

p. 51 Wellcome Library M001440; Wellcome Collection CC BY 4.0.

p. 57 CALVIN AND HOBBES © 1993 Watterson. Reprinted with permission of ANDREWS MCMEEL SYNDICATION. All rights reserved.

p. 117 Courtesy of Dr. Jennifer Gunter.

p. 139 "Welche Thiere gleichen einander am meisten: Kaninchen und Ente," *Fliegende Bätter,* October 23, 1892. University Library, Heidelberg.

p. 146 Reprinted with permission, ©1972 Dow Jones & Company, Inc. All rights reserved worldwide.

INDEX

Index—*f* preceding a page number indicates a figure; *t* indicates a table.